ADVANCE PRAISE FOR

The False Promises of Constructivist Theories of Learning

"Powerfully argued, this book provides needed insights for overcoming the loss of cultural identity, including the loss of respect for the deities, intergenerational knowledge, and nature caused by the promotion of modernist pedagogies within Andean communities."
Jorge Ishizawa of Coordinator of PRATEC (Proyecto Andino de Tecnologias)

"Western-style education is responsible for drawing children—from Mongolia to Mozambique, from Inuit to Aboriginal communities—into an urban consumer culture. Schooling effectively wipes out the self-esteem, collective memory, and ecological knowledge of cultures worldwide. C. A. Bowers' writings are an essential read for anyone seeking to understand the root causes of our eco-social crises."
Helena Norberg-Hodge, Author of Ancient Futures: Learning from Ladakh

"As exemplified by *The False Promises of Constructivist Theories of Learning*, C. A. Bowers is on the cutting edge of educational thought. Most educational thinkers continue to be trapped in the paradigm inherited from Western Enlightenment traditions, however, Bowers breaks from this paradigm and recharts the direction of global education."
Joel Spring, Author of How Educational Ideologies Are Shaping Global Society

The False Promises
of Constructivist
Theories of Learning

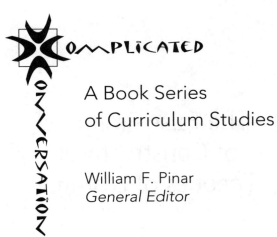

A Book Series
of Curriculum Studies

William F. Pinar
General Editor

VOLUME 14

PETER LANG
New York • Washington, D.C./Baltimore • Bern
Frankfurt am Main • Berlin • Brussels • Vienna • Oxford

C. A. Bowers

The False Promises of Constructivist Theories of Learning

A Global and Ecological Critique

PETER LANG
New York • Washington, D.C./Baltimore • Bern
Frankfurt am Main • Berlin • Brussels • Vienna • Oxford

Library of Congress Cataloging-in-Publication Data

Bowers, C. A.
The false promises of constructivist theories of learning:
a global and ecological critique / C. A. Bowers.
p. cm. — (Complicated conversation; v. 14)
Includes bibliographical references and index.
1. Learning, Psychology of. 2. Constructivism (Education). I. Title. II. Series.
LB1060.B69   370.15'23—dc22   2005006766
ISBN 0-8204-7884-9
ISSN 1534-2816

Bibliographic information published by **Die Deutsche Bibliothek**.
Die Deutsche Bibliothek lists this publication in the "Deutsche
Nationalbibliografie"; detailed bibliographic data is available
on the Internet at http://dnb.ddb.de/.

Cover design by Sophie Boorsch Appel
Cover art by Pavla Zakova-Laney.

The paper in this book meets the guidelines for permanence and durability
of the Committee on Production Guidelines for Book Longevity
of the Council of Library Resources.

© 2005 Peter Lang Publishing, Inc., New York
275 Seventh Avenue, 28th Floor, New York, NY 10001
www.peterlangusa.com

All rights reserved.
Reprint or reproduction, even partially, in all forms such as microfilm,
xerography, microfiche, microcard, and offset strictly prohibited.

Printed in the United States of America

contents

Preface — vii

Chapter 1. Introduction — 1
Chapter 2. Misconceptions Underlying Individually Centered Constructivism and Critical Inquiry — 13
Chapter 3. Toward a Culturally Grounded Theory of Learning — 31
Chapter 4. How Constructivism Undermines the Commons — 57
Chapter 5. Constructivism: The Trojan Horse of Western Imperialism — 79
Chapter 6. Toward a Culturally Informed Eco-Justice Pedagogy — 103

References — 135

preface

It became increasingly clear to me in recent years that the theories of John Dewey and Paulo Freire were not based on an understanding of the complex ways in which culture influences values, ways of thinking, behaviors, built environments, and human/nature relationships. Neither Dewey nor Freire were aware of the influence of their own cultures—particularly the way in which the language that was central to their analysis and recommendations carried forward the misconceptions and silences of the language community they were born into. Their arguments for universalizing their respective one-true approach to knowledge (experimental inquiry for Dewey and *conscientizacao*—critical inquiry—for Freire) is further evidence that they did not understand the knowledge systems of other cultures. When they did acknowledge cultural differences they relied upon a Darwinian framework whereby differences were interpreted as the expression of cultural backwardness. Dewey's lectures at the Imperial University of Japan, which were later published as *Reconstruction in Philosophy*, made no acknowledgment that his hosts had achieved anything worthwhile. Indeed, Dewey left them with the clear message that they were stuck in a spectator approach to knowledge. And when he wasn't misrepresenting the knowledge systems of other cultures in the most reductionist way, he was not reticent to refer to them as "savages." In *Education for Critical Consciousness*, Freire referred to his own more supposedly evolved way of thinking as "critically transitive consciousness," while the less-evolved ("men" who cannot think "outside their sphere of biological necessity" (p. 17), were limited by their "semi-intransitivity of consciousness"—which he equated with the existence of animals. Freire's reliance upon a Darwinian interpretative framework is also evident in his reference to oral cultures as locked into a state of "regressive illiteracy" that made it unnecessary, just as Dewey's belief that experimental inquiry made it unnecessary, to learn about the profound differences in the knowledge systems of other cultures.

Given my own efforts to base my recommendations for educational reform on a deep understanding of the differences in cultural-knowledge systems, the ways in which the languaging processes carry forward earlier ways of thinking—particularly in our culture, and the ecological impact of different knowledge systems, my criticisms in the past were focused primarily on Dewey and Freire, and on how their followers were reproducing the same combination of hubris and conceptual errors—which I continue to see as the basis of an imperialistic agenda. I was aware of Piaget's ideas, but viewed them as yet another one of the passing fads that characterizes the culture of teacher educators. It was only later, after I became aware that his ideas had become a widely held orthodoxy, that I learned that his theory of stages of cognitive development, which he called a "genetic epistemology," was yet another expression of Darwinian thinking that makes it unnecessary to learn about the differences in cultural-knowledge systems—and thus about the differences in how children learn.

This was my state of awareness when I participated for the second time in a seminar sponsored by PRATEC (Andean Project for Peasant Technologies) for members of different NGOs working in cultural affirmation projects in Peru and Bolivia. It was during the discussions of the new educational reforms being mandated by the governments of Peru and Bolivia for modernizing the schools attended by Quechua and Aymara students that I was reminded again of how Western universities have influenced the way development is understood and imposed on nonWestern cultures. But what most surprised me was learning that the educational reforms, which were to include a curriculum divided into 65 percent Western content and 35 percent local cultural content, were to be based on the core idea of constructivist learning theory—namely, that students are to construct their own knowledge. In effect, the intergenerational knowledge that has sustained the Quechua and Aymara in a wide range of ecologically challenging niches and has led to the development of one of the world's greatest diversity of edible plants, was to be replaced with the knowledge that students constructed from their encounter with the supposedly objective and scientifically based knowledge attained in the West—and from their own supposedly subjective experience. That the Western knowledge they would encounter in their classrooms has been a major factor in accelerating the degradation of the world's ecosystems and the spread of poverty did not seem an important consideration in the formulation of the educational reforms. What was important was the idea that constructivist-based learning (what is now called "transformative learning") would emancipate students from the traditional forms of knowledge that prevent them from entering the modern world.

After returning home and making further inquiries, I learned that constructivist-based educational reforms had already been initiated in countries ranging from Japan and South Africa to Islamic countries such as Pakistan and Uzbekistan. At last count, 29 nonWestern countries were introducing the theory and strategies of constructivism into their teacher-education programs and into their schools. In addition, constructivist theories of learning had long ago become the basis of teacher-education programs in English-speaking countries. The culturally transforming mix of ideas based upon a superficial understanding of the writings of Dewey, Freire, Piaget and the romantic nostalgia for recovering the child-centered phase of the earlier progressive education movement, are now being as aggressively promoted on a global scale as McDonald's and Coca-Cola. But the contribution that constructivism will make to the success of Western imperialism will have a far more destructive impact than the current spread of the West's diet of industrialized food—and the problem of obesity that is now spreading around the world.

The decision to write a book that explains how constructivist-based educational reforms represent the Trojan horse of Western imperialism led, in turn, to the realization that Dewey and Freire (and their many followers) are constructivist theorists. My earlier efforts were to explain how the key cultural assumptions that Dewey and Freire took for granted were also shared by the industrial culture they criticized. But as I pointed out in these earlier writings, the cultural assumptions they shared with the politicians and heads of transnational corporations made their criticisms little more than arguments within an extended dysfunctional family. They were correct in their criticisms of the industrial culture as a source of exploitation and nondemocratic practices. But they failed to understand the role industrial culture has played in the enclosure of the commons and in the destruction of the environment. Indeed, the shared assumptions within this dysfunctional family made it difficult to recognize that their respective liberal vocabularies were totally inadequate for understanding the diverse forms of knowledge and moral reciprocity that made the commons sites of resistance to the myth of progress that gives legitimacy to the Western model of development. For readers who have adopted the formulaic way of thinking of industrial culture as based on conservative values, I suggest they read the CATO Institute web site where it is pointed out that only in America is free-market liberalism labeled as the expression of conservatism. The web sites of the Hoover and American Enterprise Institutes also claim their allegiance to the classical liberal ideas: the progressive nature of the free-enterprise system (which depends upon critical inquiry and experimental inquiry to advance technology and exploit new markets), individual freedom (a basic misconception that ignores that human existence is part of a larger ecology of interdependent relationships), an anthropocentric way of understanding human/nature relationships, and the irrelevance (indeed, backwardness) of other cultural ways of knowing.

It was when I worked my way through the writings of Piaget and the numerous textbooks written by proponents of constructivist learning that I realized how the core assumptions of Dewey and Freire were being merged together, along with the ideas of Piaget and the jargon of current multiculturalism thinking, into a new orthodoxy for teacher educators and classroom practice. All of the god words of Dewey, Freire, and Piaget were being elevated to a level of justification that few teacher educators, and even fewer classroom teachers, had the depth of understanding to question. Constructivism, in effect, was being represented as the necessary ongoing task of reconstructing experience required for growth, democracy, emancipation, freedom, individual autonomy (which Dewey argued against), dialogue (which Freire and his followers never practice with people who disagree with them), and self-directed learning. And the silences and misconceptions that accompany the use of these god words, which often have different meanings in non-Western cultures, were also reproduced in the constructivist textbooks. I came across only one textbook that mentioned the ecological crisis. The author, however, did not understand that the various cultural commons that represent sites of resistance to the imperialism of Western industrial culture are not renewed by adopting Henry Giroux's mantra that everything should be questioned and transformed under the guidance of teachers who function as "transformative intellectuals." As these criticisms may appear too general, I would ask how many professors of education (or even certified philosophers) have passed on to their students the idea that Dewey is one of our most profound spokespersons for democracy, without recognizing that universalizing Dewey's one-true approach to knowledge is basically undemocratic—just as Freire's argument that critical inquiry is the only way that individuals can achieve the highest ontological expression of their being is also imperialist rather than democratic in the sense of not recognizing the legitimacy of other cultural approaches to acquiring knowledge and making decisions. Freire's references to the ontological project as the life force behind the drive for individual emancipation is yet another example of his failure to recognize that the mythopoetic narratives of different cultures have a far more powerful shaping influence on people's lives than the Western philosopher's abstract and culturally uninformed concept of ontology.

The problem that arises with the global spread of constructivist educational reforms is that Dewey, Freire, and the devotees of a child-centered education failed to qualify their theories by acknowledging that experimental inquiry and critical reflection are highly useful in certain situations, but that other approaches to knowledge and its intergenerational renewal might be more appropriate in other contexts—and that the contexts might vary between different cultures. In addition, their assumption that change is a constant in all aspects of life and that it will be understood the same way in all cultures as a progressive process also needed to be radically qualified. As mentioned before, a core assumption underlying the cultural imperialism of the West's industrial culture is that change is the expression of progress and that it is best achieved through the discoveries of science, technological innovation, and the drive to accumulate profits by outsourcing work to the lowest-wage earning areas of the world. This core assumption, which is also at the center of Dewey and Freire's way of thinking, makes asking the question, What needs to be conserved in light of the ecological uncertainties and the incessant pressure to transform what remains of the commons into new market opportunities—appear as reactionary?

The ability to answer the question about what needs to be conserved requires a historical understanding of the forces that continue to transform what remains of the commons into market opportunities, which at the same time undermines, as Ivan Illich pointed out years ago, the skills that enable people to participate in the mutual support systems within their communities. If acquiring this knowledge is viewed as irrelevant in a constructivist classroom and even viewed as yet another source of adult oppression, students are not likely to acquire a knowledge of such hard-won traditions as the gains made in the areas of workers' and women's rights, the separation of church and state, the right to a fair trial, the importance of an independent judiciary rather than one packed with political ideologues that support the classical liberal agenda of universalizing the idea of unrestricted market forces, and the right to privacy and free expression–even when it differs from the antidemocratic forces that are being nurtured by neoliberal think tanks such as the CATO Institute and talk show hosts such as Rush Limbaugh. And if other traditions, such as those connected with ethnic approaches to the growing and preparation of food, healing, and mentoring in the art of how to live lightly on the land, are also to be replaced by the student's own construction of knowledge—including their subjective decisions about what they want to learn—they will lack the knowledge needed to resist the changes now being promoted by the forces of industrial culture. The media is the voice of this new form of collective consciousness, and it packages its seductive message in the same language that is used to promote constructivist approaches to learning: emancipation from all traditions, greater freedom of self-expression—limited of course by the media representations of what is possible, democratic choice of consumer products, the endless stream of new technologies and improved models that are the expression of endless progress, the excitement of the transformative learning that accompanies the

Deweyian moment of doubt when the source of employment is outsourced to a foreign county—as well as the more uplifting experience of becoming dependent upon a new technology, and so forth. Dewey, Freire, and their many followers understood their guiding metaphors as contributing to progress in achieving greater social justice, but as we are now witnessing, these same metaphors can be used to create greater dependence upon a technology/consumer-dependent lifestyle—and to promote this lifestyle on a global and environmentally destructive scale.

To reiterate a key point about the weakness of a constructivist approach to learning that applies to the followers of Dewey, Freire, Piaget, and even the romantic followers of the child-centered education of the early part of the nineteenth century and the early nineteen sixties: none of them are able to ask the question, What needs to be conserved in the face of incessant technological and market driven change? Nor are they able to recognize that this question requires a combination of critical thinking, a knowledge of the traditions that are being undermined as well as how they reduce dependence upon consumerism, and an ability to assess who benefits from undermining the traditions that sustain a viable cultural and environmental commons, and who benefits from perpetuating the traditions of enclosure and oppression. Democracy requires this more complex set of understandings; unfortunately, they are not likely to be part of the education of students who are under the influence of constructivist-oriented teachers.

What neither Freire nor his many followers understand is that critical reflection may be more effective in conserving the traditions of community (and cultures) that represent resistance to the industrial culture that requires the kind of individual whose subjectively constructed knowledge alienates them from the mutual support and civil rights traditions of their community. Unfortunately, teachers in different countries of the world who have come under the influence of Western constructivist thinkers are not likely to ask about the importance of revitalizing their cultural and environmental commons as sites of resistance to the poverty and environmental degradation that follow the spread of the West's industrial culture. Indoctrinating students with the idea that they can escape what the West represents as the backwardness of their own culture by constructing their "own" knowledge reduces the possibility that they will be able to recognize which of their own cultural traditions contribute to the networks of mutual aid and non-monetized relationships and activities. Nor are they likely to be able to recognize which aspects of Western science, technology, and ways of understanding social-justice issues can be adopted without coming under the influence of new forms of colonialism.

As the above observations suggest, the globalization of the constructivist ideas of Dewey, Freire, and Piaget, which in many instances will be further reduced to slogans by educators already indoctrinated into thinking that the consumer-oriented West is more advanced than their own cultures—and thus cannot question the slogans, has implications that go far beyond the theory of learning, classroom practices, and the well-intentioned desire to contribute to a better world that characterizes the thinking of most educators. One of my aims in writing this book is to clarify the fundamental misconceptions in the thinking of the founding "fathers" of constructivist learning, to highlight how their theories of learning fail to account for the process of learning in a culture that is ecologically centered (and I am not suggesting that we should or can copy this culture or that it is the only one that could be used as a source of comparison). A third purpose in writing the book is to explain how constructivist approaches to learning lead to the form of individualism and the destruction of community that is required by the spread of the West's industrial/consumer-dependent culture. An additional purpose is to suggest that we begin to think of the teacher's responsibility in terms of mediating between the quality of life implications of adopting the practices and values of the West's industrial culture and the implications of renewing those aspects of the local cultural and environmental commons that are more ecologically sustainable and less dependent upon a money economy.

I am indebted to colleagues around the world who have helped me to understand the extent that constructivist-based reforms have been adopted in their countries—and the reaction of teachers and parents where the reforms have been in place for some time. These include Anthony T. Goncalves (Brazil), Karina Costilla, Grimaldo Rengifo, Julio Vallodolid, and Jorge Ishizawa Oba (Peru), Loyda Sanchez (Bolivia), Brooke Thomas, Levi Shmulsky, Katie Marlowe, and Caitlin Daniel (Smith College graduate students who did field research on Quechua learning patterns), Taku Sugimoto (Japan), Michael Ashley (South Africa), Kuo Shih-yu (Taiwan), Gustavo Esteva (Mexico), and Rolf Jucker and Robert Bullock (Great Britain).

While relying upon their assessment of the status of constructivist educational reforms in their respective countries, I take full responsibility for any errors of interpretation. In addition, I take full responsibility for the admittedly surface sketch of the patterns of learning of the Quechua and Aymara children that I used as a source of contrast with Piaget's genetic determinism that has been translated into the dogma for understanding the universal stages of cognitive development—that supposedly will lead (for some, as Piaget claimed) to individual autonomy. In relying upon the descriptions of Quechua and Aymara culture by people who are themselves part of these cultures, I think I got enough of it right to make the point that Piaget's theory of cognitive development might have been better grounded if he had started out as a cultural anthropologist rather than as a scientist who transferred his interest in studying the adaptive behavior of mollusks to that of children.

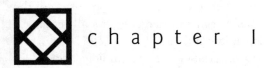# chapter 1
Introduction

The images of Western modernity are now visible in even the most remote villages and outposts of the world. Logos of companies ranging from Nike and Nokia to those of professional basketball teams are as ever present as McDonald's and KFC fast-food outlets—which are all superceded by the logos for Pepsi and Coca Cola plastered on buildings and sign boards. These logos are the surface manifestations of how Western consumerism has penetrated cultures based on profoundly different assumptions about community and the purpose of life. The globalization of Western products and services involves introducing changes into other cultures that have far more serious consequences than what accompanies wearing a baseball hat with the Nike logo, using a computer and cell phone, and eating a fast meal at McDonald's. Less visible are the Western assumptions that these products and services are based on—assumptions that underlie the connections between Western science and the drive to bring more aspects of everyday life under the control of the industrial process. These Western assumptions were also responsible for the earlier destruction of the commons as it was known in England; that is, all the natural and cultural resources that the community shared in common and that served as the basis of mutual support and moral reciprocity. These assumptions, which have placed market values and the drive to create new technologies above everything else, continue to undermine what remains of the commons in Western countries and the commons in nonWestern cultures that people rely more heavily upon for survival. What is often overlooked is that the industrial-based spread of consumerism is dependent upon a transformation in consciousness that leads to experiencing oneself as an autonomous individual. Thus, the allure of Western products needed to be supplemented in order to complete the conversion of the individual from being a participating member in an intergenerationally connected community to being a consumer-dependent individual.

The globalization of the West's view of economic and technological development is now being accompanied by the aggressive promotion of Western values and ways of thinking—through television and Hollywood films, and by Western universities that have established in the public's mind (and the minds of foreign graduates of these institutions) what constitutes high and low-status knowledge. High-status knowledge, which is represented as the basis of modernization, includes the assumption that the individual is the basic social unit, the source of intelligence and moral judgment; that literacy and other abstract forms of representation for encoding and communicating knowledge lead to a more rational and progressive mode of being; that change is the expression of progress; that Western science and technology are both culturally neutral and at the same time the highest expression of rational thought; that cultural development is governed by the laws of natural selection whereby the fittest (the more efficient and scientifically based) prevail over the less fit; and that the major challenge is to bring nature under human control and to exploit it in ways that help to expand economic markets.

These assumptions, as well as what they marginalize and subvert, are seldom made explicit when people in different cultures become consumers of Western products. But these assumptions are part of the background message that has a cumulative effect on consciousness and relationships within the different communities that make up the world's cultures. To eat at a McDonald's, to wear Levi's jeans, and to use a cell phone communicates to others in the community a higher social status, economic standing, and the rejection of the traditions shared by other members of the community. Nor are the assumptions made explicit when Western approaches to educational reform are adopted in nonWestern cultures.

Western universities have a long history of colonizing the governing elites of nonWestern countries with the same fervor and ethnocentric thinking as Christian missionaries. This ongoing process is now being supplemented by introducing Western values and patterns of thinking into the educational experience of the youth of nonWestern cultures. The Western-educated elites that control the educational bureaucracies of countries such as Peru, Bolivia, Brazil, and Mexico are now promoting what in the West is called a constructivist approach to teaching and curriculum. This same approach is being introduced into the teacher training programs in Muslim countries such as Turkey, Albania, Macedonia, and Pakistan—and in the Baltic countries, Russia, and even in South Africa and Taiwan. At last count, constructivist approaches to learning are being used as the basis of educational reform in 29 countries—and the list is growing. This same approach to teacher education is also being promoted in New Zealand, Australia, Great Britain, Canada, and the United States. Constructivism is, in fact, the dominant approach to the education of teachers in those countries that are characterized by hyperconsumerism. The connections between the form of individualism fostered through a constructivist approach to education and the process of colonizing other cultures to the patterns of a consumer-dependent lifestyle will be one of the main issues to be explored in later chapters.

For now it is important to identify the higher values and patterns of thinking that are supposedly fostered by constructivist approaches to learning and how these values and patterns of thinking are being represented as essential to eliminating cultural backwardness and non-democratic practices—which to the casual observer appear as worthy goals to be attained. The higher values have been articulated by John Dewey and Paulo Freire, two earlier proponents of constructivist based educational reforms. A more recent articulation of these highest values can be found on the web site of an organization that calls itself the Reading and Writing for Critical Thinking Project. This organization sends Canadian and American professors of education to foreign countries for the purpose of conducting workshops for local teachers and university professors in the principles and practice of conducting constructivist learning-based classrooms. The web site, which is also intended to give worldwide exposure to constructivist ideas, lists the following guiding principles:

- **Active Learning**: an approach to learning that encourages inquiry and discovery;
- **Brainstorming**: the act of freely generating many ideas about a topic, initially without critical constraint;
- **Constructivism**: a branch of learning theory in which it is held that people create or construct meaning through acts of discovery and interpretation;
- **Critical Thinking**: having ideas that go beyond what was given; seeing new dimensions in a topic; discriminating among ideas, as in argumentative writing or debate; and
- **Cooperative Learning**: group-based instruction that usually has individual accountability. (WWW.RWCT.COM. p. 2–5)

These basic principles were articulated in a somewhat different way by Dewey who argued that education should increase the student's ability to reconstruct problematic social experiences by using the scientific method of problem solving in the context of participatory decision making within the community. Paulo Freire, who had a similar worldwide following among educational reformers, argued that critical reflection should be the basis of each generation's re-naming the world of the previous generation (1974, p. 76). Contemporary followers of these two dominant Western theorists, while framing the purpose of critical inquiry in terms of the social-justice issues of class, race, and gender, continue to restate through the use of a somewhat different vocabulary the basic ideas and background assumptions that are now being promoted on a world wide basis.

One of the most important cultural assumptions underlying the high-status knowledge promoted by Western universities, by Western scientists and technologists, and by the proponents of the different approaches to a constructivist pedagogy is that change is the dominant feature of everyday life. Thus, change and progress are viewed as synonymous by most Western thinkers—which transnational corporations exploit by representing every innovation, no matter how minor or superfluous, as a progressive step over previous ones. That what is new and thus interpreted as a progressive step forward is driven by the West's incessant efforts to expand the industrialization of production and consumption does not seem to bring the myth of progress into question—except for a few critics whose voices are being drowned out by the promoters of hyperconsumerism.

The different interpretations of constructivist educational reforms reinforce the modern emphasis on change as the dominant characteristic of daily experience. One of the unfortunate consequences is that this emphasis further marginalizes the questions that should be considered as the most basic aspect of democratic decision making in any culture: namely, what traditions that support a vital and intergenerationally and interspecies-connected commons should be renewed? Except for ethnic groups, and other groups living on the margins of mainstream Western culture, the question of what needs to be conserved is seen as the expression of reactionary thinking—and even as unpatriotic. The widespread acceptance of change as being inherently progressive has deadened awareness of the many expressions of cultural traditions that are re-enacted and modified in daily life. In not being aware that most daily activities are based on traditions, the modern myth that equates change with progress becomes the dominant characteristic of how many people think. Part of the reason for thought patterns not taking account of the many traditions that are relied upon and re-enacted everyday is that public schools and universities have not provided students with the language necessary for making explicit these taken-for-granted conceptual and behavioral patterns—which is necessary if students are to assess which traditions need to be renewed and carried forward. When the educational process brings traditions into focus, they are usually the traditions that are sources of injustice and exploitation. While this is vital to the educational process, students are too often left with the impression that all traditions, except for holidays, are a source of backwardness and thus an impediment to progress.

The web sites explaining the nature of a constructivist approach to learning, as well as the educational bureaucracies of the countries it is being promoted in, emphasize that this approach will enable students to live more effectively in a world of constant change, and to be better able to help the country compete in an increasingly "survival of the fittest" world economy. When the core assumptions of constructivist pedagogies are identified it becomes easier to recognize that they support the integration of nonWestern cultures into a world economy based on Western values and thought patterns. The colonizing nature of constructivist pedagogies becomes clearer when its Western assumptions are compared with the patterns of thinking and values of cultural groups that still maintain that traditions are essential to the vitality and interdependence of their communities—or what can be called the "commons" (which I shall use as interchangeable with the non-market-oriented aspects of community-centered relationships, activities, and forms of knowledge that do not undermine natural systems).

Understanding constructivist pedagogies as yet another facet of Western colonization also brings into focus how this approach to educational reform will contribute to the further degradation of the earth's life supporting systems. As the rate of hyperconsumerism in Western cultures continues unabated, and third world cultures become dependent upon Western technology and consumer habits, the rate of environmental devastation will rapidly increase. We can see this process occurring in many parts of the nonWestern world. The efforts of the Chinese government to adopt the Western industrial model of production and consumption, including the same reliance on fossil fuels as the primary source of energy, will not only further stress the ecological systems they rely upon, but will add significantly to global warming. And as the Chinese increase their capacity to out-compete other technology centers of innovation and production, world tensions will be further increased. Generally overlooked in the worldwide effort to promote the Western model of development is that one of the primary driving forces is the further automation of the process of production—which not only makes the possession of skills and craft knowledge economically irrelevant but at the same time reduces the need for workers. That is, the Western technologically driven approach to progress creates the double bind where there is an ever-increasing need for bringing people into a money-based economy so they can purchase the ever-expanding list of consumer goods and services, and where there is an increasing number of unemployed or underemployed because robots, computers, and other new technologies have reduced the need for workers.

The promoters of constructivist approaches to education do not mention the culturally destructive assumptions it is based on. Nor do they engage educators in countries ranging from Russia to Peru in a discussion of how constructivist approaches to learning undermine what remains of the commons in these different cultures and what remains of cultural diversity. They are also silent about the connections between adopting the Western approach to modernization and the destruction of the local environments. Instead, the proponents of constructivist learning invoke the god words of modernization and progress: freedom, autonomy, discovery, democracy, individualism, progress, and so forth.

The increasing visibility of Western technology, even in the hands of militants who violently oppose the spread of Western domination, makes it difficult to give serious consideration to whether these god words will actually lead to the attainment of eco-justice within different cultures and thus to a sustainable future. But there are nonviolent forms of resistance occurring in different parts of the world. Many marginalized cultures, ranging from North, Central, and South America to Southeast Asia, India, and Africa are becoming more vocal about the need to recover their own traditions of self-sufficiency and community-centered interdependence. Whether their writings and voices will lead the proponents of constructivist learning in their own governments, as well as the educational professors in Western universities, to consider the destructive nature of constructivist pedagogies is more problematic. Nevertheless, it is especially important that this process of reconsideration begin. Nonprofessional educators also need to contribute to the discussion of reforms that strengthen the commons of the world's diverse cultures and thus help to ensure a sustainable future.

This broader discussion of the culturally transforming nature of constructivist pedagogies needs to address the following issues, which are more relevant to democratic decision making than the use of democratic slogans that cloak the process of Western colonization. These issues also need to be addressed in Western countries where the need to revitalize the commons is even more acute.

Misconceptions Underlying Constructivist Theories of Learning. There is not just one theory of learning that makes the claim that students construct their own knowledge. In addition to Jean Piaget and Lev Semenovich Vygotsky who provided radically different interpretations of the student's role in the construction of knowledge, there are also the learning theories of John Dewey and Paulo Freire who have directly and indirectly supported the core assumptions of constructivist thinking. The modern Western assumptions that emphasize student-centered learning marginalize the different forms of knowledge that are the basis of community life and how they are intergenerationally renewed. The silences and misconceptions of constructivist theorists need to be made explicit and re-evaluated within the context of the larger question of whether the Western approach to development will reduce poverty and the further destruction of natural systems.

There is also a need to question the current linking of constructivist learning theory with the classroom use of computers. Computers are highly useful in many ways—in medicine, engineering, communicating, and so forth. They are not, however, a culturally neutral technology, as many people suppose. There are many forms of knowledge that cannot be digitized, and the loss of these forms of knowledge, as will be explained later, adds further to people becoming dependent upon consumerism. There are other cultural changes that accompany the spread of computers, such as the further undermining of indigenous languages and the increasing ability of governments and businesses to gather information on people's behavior. But the impact of computers on public education will be the main concern here. Computers are being promoted as essential to the reform of education on the basis that they provide access to the data needed as students construct their own knowledge. While research that focuses on whether computer-mediated learning actually improves the learning of students indicates that there are few if any gains, computer-based education is now being promoted on a worldwide basis as a necessary aspect of student-centered learning. Thus, the misconceptions that currently surround the use of computer-based learning must be examined along with the misconceptions underlying various constructivist theories of learning. The failure of teacher-education programs to help teachers understand the connections between computer-mediated learning and creating further dependence upon Western technologies, as well as how computers undermine the face-to-face, intergenerationally connected networks that sustain the commons, will also be considered.

Identifying the Forms of Knowledge that are not Acquired by Students Constructing Their Own Knowledge. What is often overlooked is how educational theorists and classroom teachers reproduce the assumptions and silences of their own professors—who repeat with only minor variations the patterns of their professors, and so on back over the generations. The generations of professors who reproduced the Western culture's anthropocentric, patriarchal, individually centered patterns of thinking are powerful examples of this process. Another pattern that has been reproduced over many generations by the gatekeepers of high-status knowledge is to represent knowledge as the outcome of an individually centered rational process—while ignoring the culturally specific assumptions this view of rationality is based on. To make the point more directly: the majority of professors of education, as well the theorists promoting a constructivist view of learning, lack an in-depth knowledge of the assumptions that underlie their own patterns of thinking as well as the knowledge systems of other cultures. The result is that the promoters of constructivist approaches to education fail to clarify for teachers, both in Western cultures as well as in cultures where constructivist assumptions are being introduced during the student's most vulnerable period of development, the importance of the complex knowledge systems that are the basis of the student's community.

To understand how constructivist approaches to educational reform are part of a global-colonizing process it is first necessary to identify the intergenerational knowledge that is undermined when the emphasis is placed on students' constructing their own knowledge. Considering what is omitted in this approach to learning also needs to take account of the intergenerational knowledge essential to participating in the non-market-oriented aspects of life in the students' communities. The promoters of constructivist approaches to learning make the argument that the interplay between critical inquiry and the construction of new ways of thinking is essential to the development of democratic societies. This argument needs to be challenged on at least two grounds: namely, whether the imposition of Western ways of thinking that undermine the traditions of non-Western cultures can be considered to be democracy in action, and whether an approach to education that alienates students from their cultural traditions prepares them to resist being manipulated by the Western media and politicians who have bought into the Western model of development.

Why Governments in Different Parts of the World are Promoting Constructivist Pedagogies as Part of Their Strategy for Development. How governments understand the relationship between constructivist-based educational reforms and economic development is becoming increasingly important as the world moves to a global economy. While the phrase "global economy" suggests equal participation in supposedly open markets, the reality is that participation is not on equal terms. Nor does the exploitation of the Southern Hemisphere by the North disappear. The technology that is the basis of the infrastructure underlying the global economy is Western in nature, and thus reproduces in the various cultural contexts the Western assumptions about the relationship of the worker to the work process. These assumptions underlie the incessant drive to create new technologies that will further reduce the need for workers.

When governments promote educational reforms that foster the myth of being an autonomous individual in a world of unending progress they are contributing to the spread of poverty and helplessness. This is because educating young people to construct their own knowledge creates a condition of alienation whereby the knowledge and skills essential to participating in the community networks of mutual aid are not acquired. Thus, the double bind. Youth are being educated to become workers in a hypertechnological and capital-intensive environment where there is no certainty of long-term employment. The false promises are now being played out in Mexico where integration into the economic system dictated by the North American Free Trade Agreement is displacing the peasants whose cultural life has been centered for hundreds of years on the growing of corn. Now with China's ability to combine even lower-wage earners with the most advanced Western technology, Mexico's manufacturing industries are beginning to disappear. In effect, the promise of economic development that is the basis of so many governmental efforts at educational reform, including that of Mexico, is giving way to the brutal realities of global competition for markets that are already flooded with consumer goods. Yet, the government of Mexico, like many governments in other third-world countries, continued to promote constructivist-based educational reforms.

Understanding How Replacing the Western Ideology Underlying Constructivist Theories of Learning with Culturally Sensitive Approaches to Educational Reform Would Contribute to Conserving Culture and Biodiversity. Like so many Western theories, including efforts to explain cultural developments by using the theory of evolution, the theoretical basis of constructivist pedagogies incorporates a number of cultural assumptions that continue to give moral legitimacy and conceptual direction to the current phase of the Industrial Revolution that is the centerpiece of globalization. These assumptions are also the basis of the liberal ideology that has its roots in the thinking of John Locke on how private property is established, a misinterpretation of Adam Smith's economic theory, an uncritical acceptance of John Stuart Mill's idea of freedom and self-creation, and Herbert Spencer's Social Darwinism. These early theorists continued the tradition of extreme ethnocentrism that was and continues to be the hallmark of Western philosophers. By ignoring the diversity of cultural-knowledge systems, as well as the wisdom that many of these knowledge systems achieved about sustainable living within the limits of local bioregions, these theorists contributed to the messianic tradition that is now being given legal status by international agreements on free trade, and by the punitive actions of the World Bank, the World Trade Organization, and the International Monetary Fund.

These cultural assumptions, as will be explained more fully in a later chapter, are taken for granted by many liberal and so-called conservative thinkers in the West. While theorists may identify themselves with one or the other of these political labels, they continue to share the same deep taken-for-granted assumptions identified earlier: change as a linear form of progress, the individual as the source of intelligence and moral judgments, the environment as a resource to be exploited by humans, and literacy (now computer-mediated thought and communication) as evidence of an advanced stage of development. Liberals and so-called conservatives differ on how they interpret the social-justice implications of these assumptions. Yet, they both share the same prejudice against cultures that value the intergenerational knowledge that is essential to the health of the commons, and that have not made economic and technological development their primary focus.

The failure of most Western liberals and so-called conservatives to question their guiding assumptions and to recognize the role of these assumptions in current efforts to force the Western model of development upon other cultures is critical to understanding the ideological roots of constructivist learning theories. That these assumptions contribute to the spread of a consumer-dependent lifestyle that is environmentally destructive makes it especially important to understand how constuctivist approaches to learning represent the Trojan Horse that governments and educational theorists have invited in under false promises.

Identifying Alternatives to Constructivist Approaches to Education in North American and Non-Western Classrooms. Constructivist pedagogies, in being based on the assumption that students learn more effectively when they construct their own knowledge, requires that teachers play the role of facilitator. This view of the teacher's role ignores that this approach to learning is a powerful form of indoctrination. It also contributes to reducing the potential intelligence of students to what they can learn from their own direct experience. What goes unrecognized in this approach is that the cultural resources of the community, which are largely excluded by this individual or peer-group-centered approach to learning, are essential to developing the talents and skills necessary to being a contributing member of the community—and essential to reducing the students' later dependency upon consumerism.

The interconnections between the deepening ecological crisis and the loss of self-sufficiency within different cultures makes it especially important that the teacher's role be understood in a radically different way. The economic and technological pressures to adopt the ecologically destructive lifestyle of the West, which are now felt in even the remotest communities, suggest that teachers should understand their role as that of a mediator. This will require a profoundly different conceptual orientation on the part of teachers, as well as greater knowledge of the deep cultural patterns that underlie modern and more tradition-centered cultures. It will also require on the part of teachers an understanding of cultural patterns that have an adverse impact on natural systems. As will be explained in a later chapter, teachers will need to understand, and thus be able to reinforce, the traditions of different cultures that strengthen the local patterns of mutual aid, self-reliance, and use of technologies that have a low impact on the natural environment. As mediator between the colonizing pressures of the West and the intergenerationally connected community, teachers will need to help students identify the positive and negative characteristics of Western technologies, to work with the mentors within the community, and to help renew the community's accumulated knowledge of how to live more lightly on the land.

The mediator role of the teacher is as much needed in Western as non-Western cultures—though cultural differences will involve fundamental differences in how teachers carry out this responsibility. Later, the commonalties and differences will be examined more closely. There is also a need to understand the characteristics of culture where teachers, regardless of their good intentions, are less able than other members of the community to revitalize intergenerational knowledge that is embedded in the moral frameworks rooted in the community's mythopoetic narratives.

As this brief overview suggests, the discussion of educational reform should not be separated from the worldwide trends that contribute to the increasing precariousness in the quality of life for billions of the world's population. In addition to environmental changes—global warming, loss of topsoil, increasing shortage of potable free water, decline in life-supporting fisheries, etc.—the world is facing more military confrontations and a widening gap in the distribution of wealth. The spread of Western technologies and lifestyle is also contributing to the loss of cultural diversity, which in turn is contributing to the loss of biodiversity. These trends, rather than the taken-for-granted status of Western Enlightenment ideals, must be kept in the foreground in any attempt to articulate the nature of educational reforms that will contribute to eco-justice within the world's cultures. These trends also need to be kept in mind in assessing the criticism that will be leveled against this book—and the continuing grip of past academic traditions of thinking will ensure that there will be many.

 chapter 2

Misconceptions Underlying Individually Centered Constructivism and Critical Inquiry

The constructivist approaches to educational reform that are being promoted in English- and non-English-speaking countries around the world have many mutually supportive sources of origin. What is especially significant, however, is that they are based on the writings of Western or Western-influenced theorists. For example, the importance that Dewey and Freire give to critical inquiry over other approaches to knowledge can be traced back to Socrates and other Western philosophers who have given us theories that either ignored or denigrated the complex and culturally varied intergenerational knowledge that was the source of food, shelter, healing, and craft knowledge they depended upon for their daily existence. Critical reflection has also been the basis of the achievements and abuses of modern science and the Industrial Revolution. It has led to important advances in the area of human rights by bringing into question traditions that exploited workers, minority groups, and women. Critical inquiry is now raising awareness about the environmentally destructive beliefs and practices that have been taken for granted for many generations. However, like the Roman god, Janus, critical reflection has two faces looking in opposite directions: one focusing on oppressive ideas and behaviors, while the other, in being a captive of the Western myth that equates change with a linear form of progress, too often overlooks the importance of protecting traditions essential to the self-reliance and civil practices of communities from the destructive impact of new ideas and technological innovations. The assumption held since the time of Socrates is that critical inquiry should lead to change, and since the time of the French Enlightenment traditions have been viewed by Western thinkers as sources of backwardness and the protection of privilege. As we shall see, this view of tradition remains central to the thinking underlying different interpretations of a constuctivist-based education.

Constructivist theories of learning have a considerably shorter lineage. Rousseau's romantic vision of the tutor's supposedly nondirective relationship with Emile attempts to represent learning as a natural process dictated by the interests and insights of the child. But the more substantive theoretical grounding of what in the early twentieth century became known as child-centered education and is now being revived as constructivist learning, can be attributed to the varied ideas of John Dewey, Jean Piaget, Paulo Freire, the current interpreters of Alfred North Whitehead's process philosophy, and of chaos theory. Each of these educational theorists explain constructivist learning (a phrase they did not all use) in somewhat different ways. They also envision different societal goals. As will be explained later, their varied understandings of the social transformative effects of constructivist learning is less important than the deep cultural assumptions they shared in common—assumptions that now align constructivist-based educational reforms with the liberal/transnational corporate projects of globalizing the Western consumer-dependent lifestyle.

An examination of the many textbooks used in the education of future teachers and professors of education reveals that the educational theorists who laid the conceptual foundations of constructivism will not be read in any depth. The field of educational studies is the one exception, as the reading of Dewey and Freire is given special attention—but in a way that does not take account of the relevance of their ideas to addressing educational issues related to the ecological crisis, the revitalization of the commons, and to other cultural ways of knowing. Most textbooks that promote a constructivist interpretation of classroom practices will contain superficial references to the various "fathers" of constructivist thinking, which leaves students with the impression that there is a broad consensus among different educational theorists. Another characteristic of these textbooks, and even of the more in-depth reading of books by Dewey and Freire, is that critical questions are not raised about the assumptions upon which their constructivist learning theories are based. Instead, the textbooks represent constructivism as essential to the student's way of learning science, mathematics, as well as other areas of the curriculum. Even the student's choice of moral values and the leadership role of teachers are to be guided by constructivist principles of learning. Given the way in which the broad field of graduate studies in education in North America has been influenced by one fad after another, it is doubtful that the professors of early-childhood education or of educational studies (the two subareas of graduate studies that are more centered on constructivist thinking) possess the conceptual background necessary to challenge the ethnocentrism and other sources of cultural misunderstanding that are reinforced as vulnerable students work their way through their professional training courses. It is also unlikely that the professors who write the textbooks and the professors who use them would be able to explain the multiple ways in which learning occurs in the dominant Western culture and in nonWestern cultures. Ironically, it is this explanation that would expose the misconceptions and false promises of constructivist theories of learning.

There is another general characteristic of how constructivist learning is being explained that needs to be mentioned before we examine the biases and silences in the thinking of the theorists often cited in constructivist textbooks. With few exceptions, the references to theorists other than Piaget who have contributed to the special vocabulary used in constructivist textbooks (e.g., autonomy, heteronomy, accommodation, assimilation, equilibrium, zone of proximal development, learning stages, and so forth) seldom lead to exploring how a more cross-culturally informed interpretation of learning would lead to challenging the central dogmas found in nearly all textbooks used in teacher-education classes. The most important of these dogmas is that "knowledge cannot be transmitted"—only discovered by the student (1998, Von Glasersfeld, pp. 23–28). This dogma is based on Piaget's way of representing constructivism as leading to individual autonomy, and the "cultural transmission model" as leading to the condition of heteronomy (that is, control from without). Lev Semyonovich Vygotsky's understanding of the cultural/linguistic basis of learning, as well as Jerome Bruner's writings on the mediating role of language, may receive a brief mention. But the deeper implications of their efforts to give culture a more central role in understanding how learning occurs are viewed as a theoretical cul-de-sac. Their suggestion that theories of learning need to be based on understanding the diversity of knowledge systems within which children develop an identity and communicative competence within their culture fails to fit the educational goal of individual autonomy.

The idea that knowledge cannot be transmitted has been a cornerstone in thinking of other influential constructivist thinkers such as John Dewey, Paulo Freire, critical pedagogy theorists, and the followers of Whitehead such as William Doll, Jr. and Donald Oliver. As we shall see later, Dewey identified intelligence as the ongoing process of reconstructing experience. The nonreconstructed experience was categorized as a habit, which he associated with the condition of being externally controlled. As he put it, "routine habits, and habit that possess us instead of our possessing them, are habits which put an end to plasticity," which he associated with growth in the ability to reconstruct experience (1916, p. 58). Freire also made the distinction between knowledge transmitted by others, which he viewed as leading to a state of heteronomy, and knowledge acquired through critical inquiry (*conscientizacao*) that leads to new forms of action. He called the former the "banking approach" to education. The latter, as he explains in *Pedagogy of the Oppressed*, involves the human act of naming the world in ways that change it. And like Dewey, Freire viewed change as the normal condition in life—the traditions of everyday life were accordingly viewed as the source of oppression.

More recent support for the idea that knowledge cannot be transmitted can be found in the writing of William Doll, Jr. who claims that educational reforms must avoid the trap of closed systems where knowledge is transferred to students. Borrowing from Whitehead's process philosophy, Dewey's emphasis on growth in reconstructing experience and recent scientific insights into the characteristics of self-organizing systems, Doll urges teachers to recognize that a system (e.g. a classroom, the experience of learning, etc.) "self-organizes when there is perturbation, problem, or disturbance—when the system is unsettled and needs to resettle, to continue functioning" (1993, p. 163). The teacher's role is therefore that of the catalyst who introduces disequilibrium that triggers a new level of self-organization. What is especially interesting is that Doll assumes that the new level of self-organization is always an advance over what previously existed—an assumption that bears no connection to historical events. Donald Oliver, also a follower of Dewey and Whitehead, urges teachers to create a condition that fosters on the part of students the experience of "becoming." "The creative autonomy of the student," as he put it, " is impelled from within; coming from within, it has its own motive, its ideal of itself; its subjective aim" (1989, p. 117).

If theorists such as Dewey, Freire, and their many followers who now occupy professorships in English-speaking colleges of education are unable to recognize the absurdity of the idea that knowledge cannot be transferred, how can their students be expected to adopt a different way of thinking? Foreign students who may not have an in-depth knowledge of Western traditions of thinking that have influenced these educational theorists and who may still be struggling with the subtleties of the English language while at the same time being enthralled by the material manifestations of modernity, are even less likely to speak up in class when they hear that a constructivist approach to learning avoids the backwardness of the cultural transmission model of learning. Not only do the leading educational theorists share the same assumption about associating autonomy with students constructing their own knowledge, but this assumption is reinforced in the other university classes where professors represent language as a neutral conduit through which objective information is communicated to students—who are, in turn, expected to express their *own* ideas. What is most surprising is that the professors who promote the idea that knowledge cannot be transmitted fail to recognize that their own education, as well as that of their own students, represents powerful examples of what this modern educational dogma denies.

Associating individual autonomy with being better adapted to a world of constant change is one of the hallmarks of the different constructivist theories of learning that is now being promoted on a worldwide basis. But it is also important to understand more fully why the more influential constructivist theorists share the same silences. The following overview of the ideas (and misconceptions) of Dewey, Piaget, and Freire will help clarify the sources of their misconceptions that are reproduced in most constructivist textbooks. These misconceptions, as I will argue, are the main reason their ideas are unsuited for guiding educational reforms in both Western and nonWestern cultures.

⊠ John Dewey (1859–1952)

The ideas of John Dewey address such a broad range of philosophical, social, aesthetic, and educational issues that it is impossible in a few short pages to accurately represent them all. However, it is possible to highlight those aspects of his thinking that have contributed to the widespread assumptions and silences that have influenced constructivist-based educational reforms being adopted here and abroad. Dewey has always been acclaimed as an untiring supporter of democracy, the scientific method of inquiry, and the public school as an agent of social change. In effect, he became an icon of progressive educational reforms, and he is even being taken seriously again by some American philosophers—including environmentally oriented philosophers (Bowers, 2003a). But this continued and in some quarters renewed interest in Dewey's ideas ignores the silences and misconceptions that are particularly important to judging the relevance of his thinking today.

The silences and misconceptions can be organized into three categories: his failure to recognize the world's culturally diverse knowledge systems, his failure to recognize how different knowledge systems are based on intergenerational traditions that contribute to patterns of mutual aid that have a smaller ecological footprint, and his failure to recognize the need to conserve cultural traditions that represent alternatives to the industrial model of production and consumption that is now being promoted on a global basis. Some readers will argue that it is unfair to criticize Dewey on these grounds, as awareness of the interdependence between human behavior and the degradation of the environment did not really take root until the publication of Aldo Leopold's "land ethic" in *A Sand County Almanac* (1949). In dismissing my criticisms, it should be kept in mind that in Dewey's early years the indigenous cultures spread across North America were being decimated, that Henry David Thoreau and John Muir were among the best-selling authors of his day (which extended over many decades), and that a number of species, most notably an estimated 50 million bison, were being killed off to the point of near extinction—and that this was widely reported in the newspapers. There is also the problem of Dewey's current admirers who continue to overlook Dewey's silences as well as his role as an apostle of modern progress, even as environmental awareness is spreading to churches, some segments of the general public, and even to a few corporations. I will use these silences and misconceptions as the basis for arguing that educational reforms based on Dewey's thinking will exacerbate the pressures leading to the loss of cultural diversity and to the further degradation of the environment.

Dewey's definition of the purpose of education is a good place to begin, particularly since it is based on the same assumption that today's proponents of constructivist learning take for granted. Education, as he put it, "is that reconstruction or reorganization of experience which adds to the meaning of experience, or increases ability to direct the course of subsequent experience" (1916, pp. 89–90). As he states elsewhere, "the educational process is one of continual reorganizing, reconstructing, transforming" (p. 59). Traditions, which he also refers to as habits, are only relevant if they facilitate this ongoing process of reconstructing the collective experience of the community. That is, the way in which the current generation understands what is problematic and thus in need of being transformed so that it is no longer experienced as problematic determines which traditions will be taken into account in formulating a new plan of action. The wisdom of elders or of people who have been mentored to carry forward traditions of knowledge gained and refined over generations of collective experiences, such as the knowledge of the healing properties of plants, simply has no place in Dewey's way of thinking. Rather, what he places an emphasis on is that experimental knowledge must meet the test of only one generation's experience, which he sees as an ongoing process of reconstruction within the experience of that generation.

Dewey's way of equating intelligence with the utilization of the scientific mode of inquiry, where a hypothesis or plan of action has to be tested in the reconstruction of problematic situations, led him to view traditions (habits) as impediments to intelligent, experimentally oriented behavior. "Routine habits," he wrote, "are unthinking habits; 'bad' habits are habits so severed from reason that they are opposed to the conclusions of conscious deliberation and decision" (1916, p. 58). And in *The Quest for Certainty* (1960), Dewey continues to represent cultural habits as impediments to experimental inquiry. His opposition is clearly stated in the following quotation: "Knowledge which is merely a reduplication of ideas of what exists already in the world may afford us the satisfaction of a photograph, but that is all" (p. 137). Even moral values are to be continually reconstructed through the use of the scientific method of inquiry. That is, moral values regarded as intergenerationally sanctioned guides to behavior were viewed by Dewey as limiting the exercise of an experimental approach that would enable each generation to test within the context of their own experience which moral values they should live by.

Dewey's ideal of a community that comes together to make decisions based on the scientific mode of problem solving influenced the mission he assigns to the public schools. The overall purpose of schools is to socialize students from different cultural backgrounds to adopt the method of democratic, community-centered decision making. In order to achieve what Dewey envisioned as the eventual withering away of the public school, which he saw as creating an artificial separation between what was learned in the classroom and the life of the community, he urged teachers to determine which aspects of the community's traditions were outmoded and thus obstacles to the transition to a democratic, scientifically based community. As he put it, the task of teachers is to create in the classroom "a purified medium of action" (1916, p. 24); that is, omitting from the student's learning experience any encounter with cultural traditions that could not be tested and reconstructed on the basis of experimental inquiry.

To Western thinkers who shared the Enlightenment's assumptions about the oppressive nature of traditions, as well as assumptions about the progressive nature of change and the superiority of the scientific method over other ways of knowing, Dewey's ideas were beyond criticism. Indeed, his past and current followers view him as a leading thinker on how education can lead to more democratic communities. Currently, a few environmentally oriented philosophers are trying to establish that his ideas can become the basis of a more environmentally responsible lifestyle (Light and Katz, 1996). However, when we use a different set of references, such as a world of diverse cultural ways of knowing and a world where Dewey's scientific method is being used in the service of replacing traditions of community self-sufficiency with new market-oriented technologies, Dewey's ideas can be understood in quite a different light. We can recognize that this untiring advocate of democracy argued for the universal adoption of the scientific method of inquiry and the abandonment of what he called the spectator approach to knowledge. He also advocated that his understanding of a progressive linear form of change be universally adopted, and that each generation should rely upon their immediate experience in determining which traditions were relevant to reconstructing current problematic situations. Neither Dewey nor his followers recognized the contradiction in suggesting that other cultures could become democratic only as they gave up their traditions of decision making and approaches to knowledge and values.

In assessing Dewey's ideas within the context of multiple cultural knowledge and moral systems, it is important to keep in mind that he relied upon an evolutionary interpretative framework that led him to refer to cultures that were not based on his method of experimental inquiry as backward and as examples of "savages." In his way of thinking these less evolved cultures were driven by the quest for certainty, and relied upon a spectator approach to knowledge. The more evolved cultures, by way of contrast, understand "ideas are anticipatory plans of action and designs which take effect in concrete reconstructions of antecedent conditions of existence" (1960, pp. 166–167). While Dewey shared the silences of his generation toward the widespread practice of reducing the environment to an exploitable resource, his followers continue to perpetuate the progressive mind-set that is still unable to recognize that it is the intergenerational knowledge of different cultures that enables them to live less environmentally destructive lives. Thus, even though some environmentally oriented philosophers and educational theorists are attempting to represent Dewey as an early environmental thinker, his evolutionary way of explaining what constitutes backward and progressive cultures continues to come into play in discussions about what can be learned from ecologically centered cultures. Discussions usually are cut short by the comment that we cannot go back to an earlier and more primitive way of existence—a comment that indicates the continuing influence of an evolutionary way of thinking. Or the person who suggests that some forms of intergenerational knowledge are sources of resistance to the spread of the industrial-based approach to production and consumption is labeled as a reactionary thinker. The irony today is that the constructionist approach of Dewey and his current followers is actually the reactionary position. That is, it represents an attempt to go back to a way of thinking that is now being widely recognized as contributing to an industrial/market-oriented culture that is, for all its benefits, the source of cultural imperialism and environmental devastation.

◊ Jean Piaget (1896–1980)

Spread across my writing desk are nine textbooks used in the "training" of teachers. The titles of these texts reflect the different ways in which constructivism is being promoted as having the answers to unresolved educational problems. They include *Developing Constuctivist Early Childhood Curriculum* (2002), *The Constructivist Leader* (2002), *The Young Child as Scientist* (1991), *Moral Classrooms, Moral Children: Creating a Constructivist Atmosphere in Early Education* (1994). Several of the textbooks contain brief references to Vygostky's insights into the social basis of learning, and one even mentions Gregory Bateson's view of intelligence as part of the larger ecology of Mind. But the authors fail to go much beyond borrowing a phrase or two from these two seminal thinkers. The effect is to create the impression that there is a wide consensus on the constructivist way of interpreting intelligence; that is, a wide consensus that supports Piaget's theory of stages of cognitive development. The curricular and pedagogical recommendations of these textbooks also reinforce the central dogmas of Piaget's legacy that Dewey and the progressive tradition of child-centered education had created a general receptivity for. These dogma include the idea articulated by Ernst von Glaserfeld that knowledge cannot be conveyed to another person. As he put it,

> If knowledge cannot be transmitted, but must be constructed by each student individually, this does not imply that teaching must dispense with language. It implies only that the role of language must be conceived differently. We can no longer justify the intention of conveying our ideas to receivers (as though ideas could be wrapped in little packages by means of words). Rather, we will have to speak in a way as to 'orient' students' efforts at construction. 1998, p. 27

A second dogma found in all the textbooks is that the curriculum must be appropriate to the student's stage of cognitive development. As in the case of Howard Gardner's recommendation that teachers adapt the curriculum to one of the eight forms of intelligence that a student might possess, the teacher's ability to judge which stage of cognitive development the student exhibits is highly problematic. Determining the student's stage of development becomes too often a matter of adopting Piaget's designation of the age at which different stages of development are reached for the "normal" child.

The third dogma, which is actually the overall goal of constuctivist-based education, is that aligning the curriculum to the student's stage of cognitive development will lead to the autonomous individual. And the fourth dogma is that critical inquiry and experimentalism should be at the center of a process-oriented learning experience. That is, the teacher must not impede the students' construction of their own knowledge by expecting them to learn about existing knowledge.

Several of these constructivist certainties have a definite Deweyian ring to them. But it is the supposed scientific studies of childhood learning that Piaget conducted over a period of 40 years at the Institut J. J. Rousseau in Geneva, Switzerland, that are cited as the basis of the constructivist pedagogy explained in the textbooks and that are now being promoted in countries around the world. What has generally been ignored by Piaget's followers is that he was educated in the natural sciences and became an expert on mollusks. This should have led his followers to ask questions about how his narrow scientific background might account for the silences and sweeping generalizations that characterize his theory of stages of cognitive development. Just as Dewey's followers have ignored the fact that the "method of intelligence" that Dewey envisioned as the worldwide basis of democratic societies did not take account of differences in cultural-knowledge systems, today's followers of Piaget continue to ignore cultural differences.

Piaget's key ideas about how learning occurs were heavily influenced by the biological insights that helped to explain the behavior of mollusks at the bottom of the lakes around Neuchatel, Switzerland. As Keiran Egan points out, one of the most important ideas that Piaget borrows from biology to explain psychological phenomena is that intelligence should be understood as a biological process of development that is genetically driven (1983, pp. 83–95). In addition to naming his theory of cognitive development a "genetic epistemology," Piaget also relied heavily upon the biological concepts of adaptation, assimilation, and accommodation. Edith Ackerman, writing in *Constructivism in Practice* (1996), observed that Piaget understood intelligence as the ability to adjust to changes in the environment. To put this another way, intelligence comes into play in the assimilation and accommodation process that leads to a new level of equilibrium between organism and the environment (pp. 27–28).

The ability of the child to accommodate (adding to the interpretative framework of understanding that accounts for variations in the environment) is, according to Piaget, dictated by the stage of the child's biological development. What is most important about this part of Piaget's theory is his claim that all children, regardless of cultural background, progress through the same stages of development—which moves from the sensori-motor stage of the infant to the logico-mathematical stage of intelligence. In between are the preoperational stage (ages 2–7) where the logico-mathematical aspects of experience are still not understood; the concrete operations stage (ages 7–11) where reasoning begins to take account of logico-mathematical concepts that can be separated from the flow of experience; and the formal operations stage (ages 11–12) where logico-mathematical concepts begin to be formed and integrated into a coherent system that enable the student to make logical deductions about hypothesis. The last stage of cognitive development involves the ability to think logically in ways that are independent of the variations that appear to the senses. This is the stage of cognitive development that Piaget associates with the achievement of autonomy, which is the goal of the educational process.

Piaget's supposedly scientifically based theory of cognitive development led Lawrence Kohlberg to use it as the basis of a theory of moral development, which had the affect of giving further legitimacy to Piaget's ideas. As a professor of education at Harvard University, Kohlberg's writings on stages of moral development were widely accepted by professors of education in North America even though he also made universal claims that anyone who had lived in another culture or had paid attention to the moral codes held by minority cultures, would have found ludicrous. Kohlberg's stages of moral development included the following: Stage 1, moral behavior is dictated by punishment; Stage 2, moral behavior is dictated by one's immediate needs; Stage 3, moral behavior is dictated by an awareness of what helps others; Stage 4 moral behavior is dictated by conformity to authority and fixed rules; Stage 5, moral behavior is dictated by agreement on moral norms and a sense of the general good; Stage 6, moral behavior is dictated by self-chosen ethical principles that meet the test of logically based universal moral principles (1981, pp. 17–18). And like Piaget, Kohlberg recognized that while the social environment may influence the rate of biological development necessary for achieving different stages of moral development, he also held to the idea that all children, regardless of culture, go through the same stages. But he believed that not all children would be able to achieve the highest level of moral reasoning, which is the counterpart of Piaget's highest level of cognitive development associated with individual autonomy. The educational implications of Kohlberg's stages of moral development were explained by two leading advocates of Piaget's ideas in the following way: teachers need to recognize that "children must construct their moral understanding from the raw material of their day-to-day social interactions" 1994, DeVries and Zan, p. 2). Later in the textbook, DeVries and Zan warn teachers against exercising too much authority over students. Without recognizing the contradiction in their own thinking, they go on to recommend that teachers reinforce cooperation among students and an experimental approach to learning. That if students really construct their own knowledge and values they might not choose either cooperation or experimental inquiry did not occur to DeVries and Zan.

As we shall see in the next chapter that takes up the ideas of Vygotsky, Peter Berger, Thomas Luckmann, as well as linguists writing on the metaphorical basis of thinking, Piaget's way of understanding the role of language is particularly important to his argument that students must construct their own knowledge—and that the stage of cognitive development is biologically determined. His view of language is also essential to maintaining, as his followers continue to claim, that knowledge cannot be transmitted from others. Egan, one of the most insightful scholars and critics of Piaget's theory of learning, notes that Piaget viewed language competence as *following* the development of cognitive structures. To quote Piaget on this critically important issue: "Language is not thought, nor is it the source or sufficient condition for thought (1983, Egan, p. 71). As will be explained in the next chapter, this is one of Piaget's most basic misconceptions—one that led him and his followers away from considering the role of language in the intergenerational renewal of the epistemologies of different cultures. Had he bothered to investigate the connections between the root metaphors embedded in the diverse mythopoetic narratives of the world's diverse cultures, his entire conceptual edifice would have come crumbling down. And the same fate would have been experienced by Kohlberg. But as it is today, scientism was in the saddle, and the followers of Piaget continue to ignore the limited boundaries within which science can operate without becoming an ideology.

Piaget continues to influence thinking about curriculum in the early grades. One area of influence is the concept of "readiness," which is derived from the way in which Piaget represented biological development as a necessary precondition for reaching different stages of cognitive development. The teacher's task is to match the curriculum to the stage the student has reached. As an aside, this approach is not too different from recent efforts to use brain research as a guide to determining the student's cognitive readiness. There is another comparison that helps clarify what is troubling about the concept of "readiness." Piaget's use (or misuse) of science to give legitimacy to his theory of cognitive stages does not provide teachers with a fail-safe scientific approach to interpreting which stage of development the student has reached. In effect, his theory leads to the same problem that is found in Howard Gardner's theory of multiple forms of intelligence—which is also based on a biology-determines-form-of-intelligence argument (which has not been scientifically proven). Gardner's theory also places on the teacher the responsibility for determining which form of intelligence the student possesses, and then to match the curriculum accordingly. The problem in terms of both theories of readiness is that the teacher's cultural biases, subjective judgment, and own lack of knowledge become critical factors in shaping curricular decisions. But both theories provide an excuse for any misjudgments on the part of the teacher, which is that student's construct their own knowledge. For teachers to take responsibility for selecting curricular content that would involve a significant intellectual encounter with different aspects of a culture's history of achievements and failures, as well as those of other cultures, would be an imposition that would thwart the students' development as autonomous individuals. The slogan heard in colleges of education that "teachers must be guides on the side and not sages on the stage" summarizes the role teachers are to play.

As suggested earlier, Piaget's ideas have largely been embraced by professors of education whose own graduate studies were heavily influenced by the ideas of Dewey and other progressive educational theories. Meeting the needs of the child, organizing curricula in ways dictated by the child's interest, and the influence that Sigmund Freud had on the idea that adults must avoid imposing their expectations on children, were all part the taken-for-granted thinking that prevailed in early-childhood education classes. And for professors in the area of educational studies, there was a more politically charged rhetoric about the need for teachers to win the students support in the reconstruction of society in ways that meet the teachers' often ethnocentric understanding of oppression. As we shall see in the following discussion of Paulo Freire as representative of another interpretation of constructivism, the emphasis was on process—with the assumption being that the teacher's role is to stimulate the process of critical inquiry that will lead to socially transformative experiences.

Piaget's way of understanding autonomy as the ultimate stage of biological/cognitive development can easily be reconciled with Dewey's idea that the individual's experience is the deciding factor in what will be considered as problematic—and thus in need of reconstruction. What the Deweyian tradition of thinking contributed to how professors of education implemented Piaget's theory of cognitive development is the idea that problem solving requires making critical reflection central to the students' construction of their own knowledge and values. The Deweyian influence can also be seen in how some constructivist textbooks now suggest that teachers should be aware of the social nature of inquiry. While there are major differences between Dewey's ideas and those of Piaget, Dewey's emphasis on the reconstruction of experience was easily translated into the Piagetian vocabulary of constructivism—with both being viewed as contributing to a linear form of progress. One of the differences that should be kept in mind is that Dewey was an advocate of participatory decision making as the basis of social intelligence, while the goal of the educational process for Piaget was the achievement of individual autonomy. The deeper implications of this difference are being generally overlooked as the promoters of constructivist approaches to learning, in their attempt to create the impression that there is a broad consensus of support, often identify with other theorists who have conflicting ideas about fundamental issues.

⊠ Paulo Freire (1926–1997)

Egan makes an observation about an assumption shared by Dewey and Piaget, and I would add Freire, that has particular relevance to the discussion of the colonizing nature of constructivist theories of learning. He notes that both Dewey and Piaget based their theories of learning on a biological model that, in turn, was predicated on the Social Darwinism of the late nineteenth century. That is, both learning theorists thought that the spontaneous, non-scientific conjectures of children about the nature of the physical world replicated the people in oral, so-called primitive cultures (2002, p. 103). As the students' biological development leads them to higher levels of cognition (logico-mathematical for Piaget, the scientific method of intelligence for Dewey), they exhibit the higher stage of mental development of the more evolutionarily advanced cultures—such as those found in the West. Freire's understanding of the stages of cognitive development is also based on this nineteenth-century interpretation of how cultures follow a linear path of development that leads from the pre-reflective magical thinking of primitive cultures to the critical reflective and democratic thinking of modern cultures.

In *Education for Critical Consciousness* (1973), Freire identifies three stages in human, cultural, and cognitive development. He labels the first stage as "semi-intransitivity of consciousness," which is followed by the higher stage of "transitivity of consciousness," with the highest and most culturally advanced stage being "critically transitivity of consciousness: Men of semi-transitive consciousness," he observed, "cannot apprehend problems situated outside their sphere of biological necessity. . . . semi-transitivity represents a near disengagement between men and their existence. . . . (they) fall prey to magical explanations because they cannot apprehend true causality" (p. 17). He identified the oral cultures in the "backward regions of Brazil" as examples of this primitive, animal-like state of existence. The highest stage of cultural evolution, which he identified with critically transitive consciousness, is "characteristic of authentically democratic regimes and corresponds to highly permeable, interrogative, restless, and dialogical forms of thought" (pp. 18–19).

That all three learning theorists share an evolutionary interpretation of how cultures develop in a straight line from primitive to modern, that is, Western, may account for why they and their contemporary followers continue to use the Western mind-set as what other cultures should aspire to achieve. That cultures have taken different pathways of development and achieved profoundly different ways of achieving complex and, in many instances, self-sustaining communities is simply ignored by the different traditions of constructivist reformers. Their proposals for educational reform, in effect, represent the imposition of the Western model of development. The traditions of knowledge and moral reciprocity that do not fit the Western, high-status model of consciousness are thus to be omitted from the constructivist curriculum. While they use the Western god words of emancipation, autonomy, progress, and so forth, they see their mission as agents in the process of cultural evolution. That natural selection fosters diversity of species, while their nineteenth century interpretation of evolution leads to the vision of a world monoculture, does not seem to matter. Their vision also includes a modern concept that is not part of nature's process of design—the idea of a linear form of progress.

Freire is important to the discussion of the misconceptions underlying the current push to impose constructivist-based educational reforms on nonWestern cultures. Freire, a Brazilian educator who has attracted a worldwide following, is more in the constructivist tradition of Dewey than that of Piaget. Like Dewey, he was concerned with social justice issues and developed a theory of educational practice that was represented—and widely accepted—as free of cultural impositions. While Dewey emphasized the scientific method of problem solving in the context of participatory decision making, Freire emphasized the need to think critically about the taken-for-granted patterns that were the basis of exploitation and general oppression that keep people from realizing their full potential as human beings. The strong social reform orientation of Freire should have led to opposition between Freirean thinkers and the followers of Piaget. Aside from Freire's criticism of Piaget's lack of a social reform agenda, what has become the Freirean tradition—now called critical pedagogy—has had the effect of supporting the general idea that students should construct their own knowledge. And as the current followers of Piaget incorporate the process of critical reflection and a concern with social-justice issues into the students' practice of constructing their own knowledge, the real differences between the two constructivist positions disappear even further.

There are a number of ways that Freire and his followers use a more social-justice-oriented vocabulary that ends up supporting the core assumptions of both Deweyian- and Piagetian-inspired interpretations of constructivist educational reforms. As mentioned earlier, Friere shared the belief of Dewey and Piaget that "knowledge cannot be transmitted." His use of the phrase "banking approach to education" is similar to Dewey's rejection of what he termed the "spectator approach to knowledge." The following statement by Freire is perhaps the most revealing of how he equates the critical reflection of each individual and of each generation with the ongoing project of emancipation from the domination of previous generations. In *Pedagogy of the Oppressed*, perhaps his most famous and original book, he writes "There is no true word that is not at the same time a praxis. Thus to speak a true word is to transform the world" (p. 75). Piaget lacked the concern with social transformation, but shared the same idea about individual autonomy even if his approach to achieving it was different in some important ways.

Another core assumption shared by different interpretations of constructivism is best articulated in the following statement by Freire.

> Human existence cannot be silent, nor can it be nourished by false words, but only by true words, with which men transform the world. To exist, humanly, is to *name* the world, to change it. Once named, the world in its turn reappears to the namers as a problem and requires of them a new *naming*. Men are not built in silence, but in word, in work, in action-reflection.
>
> But while to say the true word—which is work, which is praxis, is to transform the world, saying that word is not the privilege of some few men, but the right of every man. Consequently, no one can say a true word alone—nor can he say for another, in a prescriptive act which robs others of their words. 1974, p. 76

As the followers of Freire's ideas are likely to object that I am quoting his earliest writing and that he adopted in his later writings a more complex and culturally informed understanding of when critical reflection is highly useful, I will quote from the *Pedagogy of Freedom* (1998), which was published after his death. The central idea of *Pedagogy of the Oppressed* is also the central idea of his last book, which can be seen in his statement that "to teach is not to *transfer knowledge* but to create the possibilities for the production or construction of knowledge" (p. 30, italics in original). He goes on to claim that "the educator with a democratic vision or posture cannot avoid in his teaching praxis insisting on the critical capacity, curiosity, and *autonomy* of the learner" (p. 33, italics added). It is important to note that democracy for Freire means that the nonWestern cultures must abandon their own systems of knowledge—which is the same position that Dewey took—also in the name of achieving democracy. Freire even extends his core idea that each person must speak a true word and not live by the words of others—particularly the words of the older generation. As in all of Freire's pronouncements, there is no recognition of differences in cultural contexts, the possibility that some forms of intergenerational knowledge are essential to membership in the commons and that autonomy is an ideological construct of Western thinkers who did not (and still do not) understand how thinking always reproduces even as it individualizes the taken-for-granted cultural patterns of thinking.

The failure of Freire and by extension, his current followers, to recognize the differences in cultural-knowledge systems is now being criticized by third world activists who attempted to use his ideas as a means of combining consciousness raising with literacy programs. These activists were personally close to Freire and some held important leadership positions in international institutes dedicated to promoting Freirean-based social reforms. Siddhartha, for example, was the international coordinator of the Freirean institute in Paris (INODEP) and was the Asian INODEP coordinator for four years. Gustavo Esteva is one of the most important spokespersons and grass roots activists working on behalf of the indigenous cultures in Mexico and other regions of Mesoamerica and was an early supporter of Freire's ideas. At great personal risk, Loyda Sanchez promoted the Freirean literacy program with the peasants of Bolivia, while Grimaldo Rengifo Vasquez, who is now co-director of PRATEC in Peru, spent his early years introducing the use of Freirean literacy programs among the Quechua. They have all written about their disillusionment when they discovered that Freire's ideas required the adoption of the Western tradition of critical reflection as well as other key Western assumptions. In effect, they discovered that the use of Freirean ideas represented yet another example of colonizing the indigenous cultures to adopt the Enlightenment mind-set of the West (Bowers and Apffel-Marglin, 2005). While Freire himself wrote about the importance of dialogue and the dangers of turning his ideas into a rigid dogma, his followers were nearly successful in blocking the publication of the essays of these former advocates of the Freirean approach to emancipation.

◊ Concluding Remarks

Piaget and Freire may appear poles apart, particularly in terms of the issue of the ongoing project of individuals expressing their humanity by continually renaming the world. But they share the same assumption that their respective belief in the one true approach to knowledge should be imposed on the world's other cultures. They also shared the same case of extreme ethnocentrism by ignoring the other approaches to knowledge easily found in both Western and non-Western cultures. And like Dewey, Freire ignores that a democratic society may be guided by assumptions other than their prescriptions for a life of ongoing change and intergenerational alienation. It would not be unfair to say that the Western myth of unending progress underlies all three interpretations of constructivist-based education.

These criticisms of Dewey and Freire should not be interpreted as a wholesale rejection of the importance of critical reflection and the scientific method of inquiry. Rather, my basic criticism is that they did not recognize that the cultural context is vitally important and that when context is ignored, their respective ideas can become, like today's promoters of Piagetian ideas, part of the process of colonizing other cultures to a Western pattern of thinking and individually centered lifestyle. Their respective assumptions that they possessed the one true approach to learning brings up another aspect of colonization that is now being given attention by third world activists and some environmentalists. That is, the followers of Piaget and Freire and most of today's followers of Dewey continue to ignore the connections between their approaches to knowledge and the ecological crisis. As noted before, Dewey can be partly excused for sharing the silence of his generation on this critical issue, but his followers cannot be let off the hook so easily. Their myopia on this critically important issue has led them to ignore asking about the forms of knowledge in different cultures that contribute to living less ecologically destructive lives—and to asking how these knowledges are encoded and intergenerationally renewed.

We shall take up the educational implications of this question in later chapters when we consider the forms of intergenerational knowledge that are lost when students are indoctrinated with the idea that the only reliable knowledge is what they construct for themselves—or name for themselves (Freire) or reconstruct for themselves (Dewey). First, we need to address the complex nature of the how cultures intergenerationally reproduce themselves and the educational implications of recognizing that many aspects of this process are the basis of empowerment—if we can understand that word as also encompassing what enables people to live in mutually supportive and environmentally less destructive lives.

 chapter 3

Toward a Culturally Grounded Theory of Learning

What is surprising is the way in which Dewey, Piaget, Freire, the followers of Whitehead such as Doll and Oliver, and the critical pedagogy theorists think in terms of oppositional categories. In order to highlight the uncompromising correctness of their theories, these otherwise highly intelligent men rely upon sharply defined boundaries that have no relationship to daily experience—including their own experience. Dewey, for example, states that the opposite of experimental inquiry is the spectator approach to knowledge. Freire's oppositional categories are *conscientizacao* (roughly, critical reflection) and the banking approach to education, while Piaget sets autonomy over heteronomy. For Doll, the opposite of his "open system" way of thinking is the closed system which, ironically, he identifies with modern culture. Oliver also thinks in terms of opposites: education either contributes to "becoming" or degenerates into "transferring knowledge." Van Glasersfeld's categorical statement that "knowledge cannot be transferred" exhibits the same habit of dichotomous thinking. Underlying these diverse examples of thinking in terms of opposites is a more deeply held and similarly unexamined cultural schemata that led these theorists to think of progress as the opposite of tradition and, by extension, liberalism as the opposite of conservatism. It is interesting to note that this is a cultural pattern of thinking that duplicates mainstream Christianity's radical separation of good from evil.

What is particularly surprising about this group of constructivist theorists is that they either claim a scientific basis for their ideas, or claim that the "process" nature of nonculturally mediated experience is the basis of their insights. Yet none of their rigidly defined oppositional categories has a basis in everyday experience—including their own embedment in the culturally mediated behaviors and thoughts that get them through daily life. Equally surprising is the way generations of their followers have accepted the division of their own experience into such conceptually neat, airtight, and culturally uninformed categories.

To counter the major premise of these theorists about individuals and groups constructing their own knowledge, it is necessary to explain how the everyday reality of people is culturally constructed and sustained. This explanation also needs to account for how individuals are involved in this process and for cultural differences. In short, I will show that the "cultural transmission model" they all reject is an inescapable aspect of learning the languaging processes of the culture an infant is born into. Furthermore, I want to explain that while there are examples of cultures and even teachers that are authoritarian and destructive of human potential, they should not be used as a basis for claiming that the transmission model of culture and its implications for educators can be ignored. The basic fact of human existence is that we are all nested in a complex cultural ecology, and the culture is nested in the ecology of natural systems. And these more complex cultural and natural ecologies are an inescapable part of the classroom—even though the teacher's education has not taken them into account. The point of making the effort again to think against the grain of the modern superstitions that make constructivist theories of education so problematic is to identify how the dynamics of this intergenerational "transmission" model can be altered by teachers in ways that reinforce cultural patterns of thinking and values that contribute to achieving eco-justice within the context of different cultures.

I have deliberately used the word "transmission" in order to clarify that I am arguing from a culturally informed perspective that the constructivist theorists reject. Yet I find the metaphor, when it is associated with a sender/receiver model of communication, to be deeply problematic. Now that I have made the point that I am challenging the most basic assumptions shared by the constructivist learning theorists, I will use the phrase "intergenerational renewal" as an alternative to "transmission" in making the case that what we experience as "reality" is culturally constructed and that in varying degrees, depending upon the culture, the cultural patterns are given individualized expression that may lead to their modification. I also want to argue that intergenerational knowledge is, in many instances, a source of personal and community empowerment—and that many of the forms it takes represent sources of resistance to the spread of the Western model of industrial production and consumption. In the next chapter I will argue against a form of intergenerational renewal within the scientific community that is again using Darwin's theory of natural selection to justify the superior fitness of a computer based "global intelligence" over the diversity of cultural-knowledge systems.

There is a way of determining whether the sharing of cultural knowledge is an inescapable part of human existence or whether it can be avoided by implementing one of a variety of constructivist approaches to learning. The first way would be to examine why the various constructivist theorists share so many cultural patterns that they have personally individualized in only minor ways. That is, can their own construction of knowledge be used to account for the patterns they share with other people who are socialized to write from left to right, use standard spellings, paragraphs, and capitalizations? Did all of the constructivist learning theorists individually originate the subject-verb-object pattern used by other speakers of English? Questions could also be asked about how, if knowledge is individually constructed, they share similar beliefs about a trial by a jury of peers, the need to stop exploiting workers, and the meaning of street signs. The list of shared cultural patterns could be vastly expanded, but these everyday examples of shared knowledge are sufficient for asking the question, Why do constructivist learning theorists present an explanation of how children learn that does not account for the tacit and explicit forms of knowledge they base their own lives on? As the classroom practice of their theories can have an important influence on the life chances of students, they and the teachers who try to implement their ideas should be held to a minimal level of accountability: namely, that their theories accurately account for how students learn to be communicatively competent within the context of their own cultures.

Discovering that theorists seldom live by the theory they want other people to base their lives on should not come as a great surprise. But there is another reason for challenging the theoretical underpinnings of constructivist-based educational reforms now being promoted in countries around the world. If the constructivists are correct in equating the cultural transmission model with the spectator approach to knowledge (Dewey), with the banking approach to education (Freire), and the closed systems that lead to minds closed to the possibility of "becoming" (Doll and Oliver), then there is no reason for classroom teachers to be knowledgeable about the intergenerational knowledge that sustains the life of the surrounding communities. Indeed, learning about the traditions of the different cultural groups would be a waste of time for teachers who hold to the idea that students construct their own knowledge and values. Integrating the intergenerational knowledge of the community into the curriculum would be tantamount, according to any one of the constructivist theories, to undermining the students' ability to move from a condition of cultural backwardness into what educational theorists like to refer to as the postmodern era. In addition to the misconceptions that underlie the ideas of all of the constructivist learning theorists and their interpreters in colleges of education, their view of other cultures represents another example of the hubris of Western thinkers who view their mission as that of saving nonWestern cultures from their backward and oppressive traditions.

The contradiction that is at the center of constructivist theories of learning, where the constructivist teacher's approach to multicultural education largely takes the form of teaching tolerance while at the same time ignoring the importance of helping students acquire an in-depth understanding of the diversity of intergenerationally based-knowledge systems, can partly be explained in terms of the role that language plays in influencing awareness and what will be ignored. Aside from Vygotsky's understanding of how language influences thought and the formation of individual identity—Vygotsky was aware of George H. Mead's symbolic interactionism—all of the constructivist theories had either a totally wrong understanding of language (Piaget, Doll, Oliver), or they viewed its role as reproducing old and thus oppressive ways of thinking (Dewey, Freire). In *Knowing and the Known* (1949, Dewey and Bentley), Dewey explains how language can get in the way of experimental inquiry in the following way: "The naming of the observation and naming adopted is to promote further observation and naming which in turn will advance and improve." He goes on to say that "This condition excludes all (previous) namings that are asserted to give, or that claim to be, finished reports on 'reality'" (p. 49). And we have only to recall Freire's statement that "to speak a true word is to transform the world" to recognize how he, like Dewey, viewed language as one of the chief impediments to critical reflection and social change. What can be said on Freire's behalf is that he had a partial understanding of the political nature of language.

If we are to develop an alternative to the constructivist theories of education that contributes both to the colonization of other cultures and to undermining the sources of resistance to the spread of the industrial approach to production and consumption that is having such a devastating impact on the environment, we will need to clarify the many ways in which cultures reproduce and renew themselves. Ironically, even though the constructivist theorists, as well as the professors of education and classroom teachers, can never exist apart from the cultural ecology of their time and place, their theories largely ignore the multiple ways in which thought, behavior, and identities are influenced by culture. As an alternative to thinking of the teacher's role as facilitating the reconstruction of experience, emancipation from the oppression of the previous generation, becoming, and the attainment of autonomy through logico-mathematical thinking, I would like to suggest that we begin to think of the teacher's role as a mediator in the process of what Jorge Ishizawa calls, intercultural and, I will add, intergenerational renewal. Unlike the constructivist teacher whose primary mission is to facilitate change, the mediator's role requires a profoundly different orientation. That is, the mediator needs to be knowledgeable about the culturally different forms of intergenerational knowledge, which is different from prejudging them as inherently backward and oppressive. And the teacher needs to be knowledgeable about the many ways in which the languaging processes in different cultures contribute to forms of intergenerational renewal essential to ecological sustainability, as well as to community and environmentally destructive patterns. That is, in adopting the role of mediator, the teacher has to be open to the possibility that traditions can be sources of empowerment just as they can be sources of exploitation and ignorance. Like the physician who needs to understand human anatomy and the lawyer who needs to understand the foundations of the law, teachers need to understand the cultural ecology that influences their ideas, values, and every aspect of classroom communication, as well as the cultural ecology of their students. That is, at the core of their professional knowledge should be a deep understanding of culture in all its varied dimensions. In the next section I will discuss aspects of cultural renewal that have particular relevance to understanding the teacher as an intercultural and intergenerational mediator.

⊠ The Cultural Construction of Knowledge and Personal Identity

If this were a chapter on the history of theorists who wrote about the influence of language and patterns of social interaction on the child's thought, behavior, and self-identity, it would be necessary to start with the insights of Vygotsky, the Russian linguist and psychologist. The American who contributed much to the field of symbolic interactionism, George Herbert Mead, would also be included. But I will skip over their contributions in order to focus on more recent explanations of the social (I prefer cultural) construction of what people in different cultures take to be everyday "reality." The writings of Peter Berger, Thomas Luckmann, Mark Johnson, George Lakoff, Richard Brown, Edward Shils, and Gregory Bateson can more easily be translated into culturally informed pedagogical practices. Of the above, only Vygotsky is occasionally mentioned in constructivist textbooks. As the current understanding of the constitutive role of language in the formation of different forms of cultural intelligence has advanced well beyond Vygotsky's pioneering work, I will focus here on the most recent insights.

There is a special need to highlight the role of language in facilitating intergenerational renewal that has significance for teachers who are making curricular decisions that contribute to lifestyles that are exacerbating the ecological crisis. Contrary to the thinking of the constructivist theorists, the primary issues that need to be addressed have to do with a rapidly degraded environment that further contributes to the spread of poverty, the maldistribution of wealth that leads to patterns of hyperconsumption for a few and disrupted local economies for the many, and the loss of intergenerational knowledge that represents, in many instances, sources of resistance to the globalization of a consumer-dependent lifestyle.

By identifying the characteristics of intergenerational renewal that are shared by all cultures, I hope to clarify the more limited yet essential role that critical reflection can play in the process of intergenerational renewal. I also hope to clarify how the dominant beliefs of a culture can be based on the assumption that the present practices should fit the rigid and unchanging prescriptions borrowed from the past. While many people want to cite these examples as expressions of conservatism, it would be more accurate to call them examples of reactionary thinking. From time to time, I will refer to the specific beliefs and practices of different cultures in order to ground my explanations. For the most part, however, I will explain different aspects of the cultural construction and renewal processes that have particular relevance for rethinking the radically reductionist argument that critical thinking is the *only* way in which knowledge is acquired and reforms achieved. The starting place is to challenge what seems to be the linchpin in the various interpretations of constructivist approaches to education: namely, that knowledge cannot be transferred, but can only be constructed by the student through critical reflection, the experimental method, by becoming, and by achieving a certain stage of biological/cognitive/moral development.

If professors of education had taken seriously the Berger and Luckmann tradition of the sociology of knowledge, rather than the Marxist tradition that captivated the thinking of many educators from the early seventies until fairly recently, they would have found a more adequate explanation of the role of language in reproducing a culture's stock of knowledge. More importantly in terms of accounting for the silences shared by constructivist theorists about the nature of culture, Berger and Luckmann explain why individuals can re-enact, even individualize, cultural traditions without being explicitly aware of them as cultural traditions. In summarizing key insights about the constitutive role of language I will draw on the insights of other theorists as well.

Unlike Piaget's claim that language competence follows the development of cognitive structures and Freire's claim that individuals must avoid the false words of previous generations by naming the world themselves, Berger and Luckmann point out that at the center of the relationships that constitute the individual's social (cultural) experiences are the multiple processes of communication. These include the non-verbal exchanges, the spoken and written word, thought patterns encoded in and communicated through the material/built culture (design of buildings, roadways, organization of public spaces, clothes, and so forth). In some cultures the plants, animals, rocks, wind, rivers, etc. also are understood as sources of communication. In effect, Berger and Luckmann represent the languaging processes of a culture as its core constituting and sustaining characteristic. As they put it in *The Social Construction of Reality* (1967), a book that should be part of the teacher's professional studies, "Everyday life is, above all, life with and by means of the language I share with my fellowman. An understanding of language," they continue, "is thus essential for an understanding of everyday life" (p. 37). As indicated above, what they refer to as language should be understood as the multiple languaging processes that encompass every form of communication—which will vary from culture to culture. Colors, designs, smells, sounds, bodily gestures, organization of physical space, doorways, building materials, clothes, different foods, and so forth are all part of a culture's languaging processes that constitute, guide, repress, and transform the ordinary into the extraordinary, and so forth. In short, they are all part of the complex ecology of symbols that the members of the culture interact with as sources of meaning, status relationships, and identity formation.

The Berger and Luckmann explanation of the cultural construction, renewal, and reification also accounts for how members of a culture can exhibit communicative competence (in the anthropological meaning of the phrase) without being explicitly aware of the shared cultural patterns they are reenacting—and in some instances giving individualized interpretations that do not fundamentally change the patterns. They explain that this cultural knowledge is largely taken for granted; that is, the person has a natural attitude toward the cultural patterns that have not been named and thus made explicit. Thus, the individual has a natural attitude toward the cultural patterns of metacommunication with others (maintaining eye contact, using tone of voice, other bodily gestures—that are shared by other members of the culture), as well as other cultural patterns such as the design of dwellings, how to relate to the environment and people regarded by other members of the culture as inferior, and even ideologies that represent possibilities that cannot be lived. In contrast to the taken-for-granted nature of most of our cultural knowledge, the constructivist learning theorists as well as most teachers from the elementary grades through graduate classes emphasize the student's explicit knowledge—that is, what they can reflect on and articulate in spoken and written form. The aspects of taken-for-granted knowledge that are explicit are often associated with some social-injustice issue, such as discriminatory hiring practices, racial and gender prejudices, misunderstandings that have been passed on from generation to generation, and so forth. What is seldom made explicit are the taken-for-granted patterns that are sources of empowerment, that contribute to non-violent ways of settling disputes, that carry forward the important achievements of the past—such as the gains made in the protection of workers' rights (now being eroded), and so forth.

Even when specific taken-for-granted patterns are made explicit in the process of critical reflection and changed, other taken-for-granted beliefs remain unexamined. For example, recent efforts to make explicit racist and sexist cultural patterns did not lead to a critical examination of other cultural patterns, such as those based on the assumption that change is a linear form of progress, that consumerism is the highest expression of personal success, and that the pursuit of individual self-interest is a universal value that should be imposed on other cultures. Instead, the achievement of a non-racist and non-sexist society was supposed to enable everyone to pursue the American dream of material success—which is overwhelming the natural systems we depend upon with needless waste.

By ignoring the extent that taken-for-granted cultural patterns are an integral part of daily experience, the constructivist learning theorists fail to recognize the limits of critical inquiry and the individual's ability to construct their own knowledge. The simple fact is that if a cultural pattern of thinking is taken for granted, such as Dewey's idea that democracy requires displacing other cultural ways of knowing with the Western model of scientific inquiry or Freire's idea that all the world's cultures should adopt critical inquiry as their *only* approach to knowledge, it will not be made explicit and thus be open to critical reflection. Rather, the deep and largely unconscious patterns of thinking, which in the above examples imply that only one legitimate approach to emancipation that must be universalized, provide the conceptual framework within which critical inquiry can be exercised. To make the point more succinctly, critical reflection is always based on a culturally specific set of taken-for-granted assumptions and patterns of behavior.

As the promoters of high-status knowledge have emphasized that everyday life should be guided by rational thought, which the constructivist theorists have narrowed down to critical reflection, something more should be said about the fact that not all taken-for-granted cultural patterns contribute to social injustices. Nor do they all limit the potential of individuals. As mentioned above, the taken-for-granted cultural patterns that lead some groups to protest the impact of these patterns on their lives are the ones given attention in public school and university classrooms, and the taken-for-granted patterns that should be renewed in the lives of students are generally ignored. I suspect that this statement will cause many readers to charge me with being a reactionary thinker—which would be a mild label compared to one of my critics who suggested that I am a neofascist because I write about environmental issues.

⬖ Toward a More Complex and Balanced Understanding of Tradition

Taken-for-granted cultural patterns can also be called traditions, which is another word (iconic metaphor) that still encodes the analogies that were the basis of how French Enlightenment thinkers understood the nature of tradition. That is, they viewed the church, the remnants of the feudal system, and the folk superstitions of their day as impeding the emergence of a progressive form of society based on rational thought. Today, most cultural mainstream teachers and professors either associate the word *tradition* with holidays, and more broadly with ignorance and cultural backwardness. It is common to hear "traditional" used to refer to cultures that are considered as undeveloped, backward, and based on "pre-scientific thinking" (i.e., superstitions) to quote E. O. Wilson and the late Carl Sagan. For the last 15 years or so I have urged educational reformers to recognize that the word *tradition* is as inclusive as the word *culture*. I have further argued that representing tradition as the opposite of progress is to support the taken-for-granted assumptions underlying the Industrial Revolution—which is itself dependent upon carrying forward and modifying traditions in the areas of science and the development of technology. But the tradition of thinking of tradition as a backward cultural practice and pattern or as an appeal for the return to the traditions that privileged certain groups over others is so taken for granted that it is beyond the scope of critical reflection. I would be surprised to learn of any educational theorists who followed up my suggestion that Edward Shils' book, *Tradition* (1981), should be essential for classroom teachers and anyone else who thinks about educational and environmental issues. There are many ironies connected with the misconceptions that educators have about the nature of traditions—misconceptions that in any other profession would be grounds for the charge of malpractice. One of the ironies that still is ignored is that in turning "multicultural education" into a mantra, while ignoring the traditions that are the basis of different cultural (ethnic) groups, is to engage in a fundamental contradiction. A cultural (ethnic) group is distinguishable from other cultures by virtue of the fact that, in addition to the traditions it shares with the rest of the cultures that have come under the sway of modern culture, it has it own distinct traditions.

As reductionist thinking usually takes over when the word *tradition* is mentioned, it will be useful to summarize the main points again: namely that a theory of learning needs to take account of the complex nature of traditions. This includes an awareness of the many sources of individual empowerment, technological competency, community self-sufficiency, and civil liberties. Depending on the culture, other traditions may be the source of inequality in terms of political decision making, educational opportunities, the legal system, and the distribution of wealth. And most of our high-status traditions that drive economic and technological development are still based on earlier assumptions that represented the environment as an exploitable resource—even though these developments are identified as the latest expression of progress. If we take seriously Shils' explanation that traditions include every aspect of culture that has been re-enacted (and even modified) over four cohorts or generations, then it is possible to recognize that the constructivist claim that "knowledge cannot be transferred" is an example of abstract thinking that has no basis in everyday reality. Taking account of the traditions that were relied upon in representing this idea in print points to two serious problems: that the author of the statement does not know how to use language in an accountable way, and that so many constructivist professors would take him seriously. Shils is not (nor am I) making the case that we should accept all traditions in their present state of development. Rather, he is saying that traditions are an inescapable part of human existence. He is also saying that we should be aware of the characteristics of traditions in our pursuit of the French Enlightenment ideal of freeing ourselves from the hold of tradition. Shils refers to this tradition of thinking, which is shared not only by constructivist learning theorists but also by Marxists, scientists, and classical liberal thinkers we mistakenly call conservatives, as an "anti-tradition tradition." That is, this taken-for-granted pattern of thinking has been carried forward over four generations.

Western thinkers, nearly to a "man," have maintained a long tradition of representing rationality and critical reflection as the source of knowledge. In the process they have ignored how much of daily life is based on taken-for-granted traditions—as well as traditions that have been consciously chosen as worthy of being continued. When Shils argues that traditions are everything handed down and reenacted over four generations he means everything from the use of standard spellings, the many technologies that go into the production of a book, the value and use of currency, the design and technology used in building and flying an airplane, the use of spices and recipes, health care practices, the multiple-layered procedures that govern elections and the legal system, the structure of a narrative and a play, the traditions of music—from folk, jazz, to classical, and so forth. Understanding the nature of traditions becomes even more necessary and complicated when we take account of how different cultural mythopoetic narratives lead to fundamental differences in the traditions that are taken for granted by the members of other cultures.

By ignoring how human life involves re-enacting and giving individualized interpretations of traditions—and even rejecting some traditions, from the time of conception and birth to the end of life (which some cultures view as the beginning of a new cycle of life), constructivist learning theorists and classroom teachers are passing on to students the modern tradition of thinking that we can live better if we don't have any traditions—except for holidays and religious observances. The message conveyed to students is that they can create their own traditions—which, in turn, are not to be imposed on the following generations. This will allow them to experience the excitement of "becoming" in a hyper-media environment where new images and products are constant reminders that progress is unending—as long they do not exceed the limit of their credit card or lose their job to corporate outsourcing and greed.

Shils' observations about the nature of tradition, which should not be confused with specific traditions that vary from culture to culture, include the following: (1) that traditions do not pass themselves on, but are perpetuated by human beings; (2) that traditions are not static, but like a plant undergo constant change—though a tradition of belief (what Shils refers to as "traditonalism") may hold that traditions should not change or that we should go back to earlier traditions; (3) that some traditions change too slowly while other traditions may be overwhelmed by the development of other traditions before people are aware of their importance (such as the tradition of privacy being undermined by further developments in the traditions that underlie computer technologies); (4) that when a tradition is lost it cannot be recovered—though a somewhat different tradition may eventually take root; and (5) that some traditions should not have been started in the first place—and that the traditions of critical reflection and democratic decision making are essential in overturning them. By using the plant as an analogy for understanding the nature of traditions, with its new growth amid dying roots and branches, Shils is emphasizing that traditions should be regarded as organic and thus undergoing a constant process of development—which requires both constant pruning and nourishment.

Shils' understanding of tradition represents another way of understanding the historical continuities within a culture—or what I refer to as an intergenerationally connected culture. It is also interesting in another sense that has more relevance for teachers than any other profession. His view of tradition, which he describes in great detail rather than to argue for the creation of traditions, is similar to Edmund Burke's recommendation that changes should be assessed in terms of making a constructive contribution to the well being of the community. While other aspects of Burke's political philosophy supported traditions that were wrong for his day, and even more wrong in our times, he understood that critical reflection needs to be balanced by a sense of responsibility for carrying forward the genuine achievements of the past, and for ensuring that the adoption of current changes do not diminish the prospects of future generations. Today, additional criteria for assessing the worth of innovations should include the impact on the traditions of economic self-sufficiency of communities within the West and third world cultures, and the impact on the viability of the world's ecosystems. Simply put, the tension between the forces of change and living traditions calls for the exercise of critical reflection—but critical reflection that is not based on the assumption that change automatically leads to progress.

In terms of technological innovations, this more balanced and sane use of critical reflection might lead to asking whether a new technology makes obsolete the craft knowledge and skills of the worker, how it alters the relationships between workers, whether it reduces workers to performing a segmented role in the production process, who benefits economically from the innovation and whether it makes the economic viability of the community more precarious, and what is the impact on the environment. This approach to using critical reflection does not assume that the traditions of the community must be replaced, but rather is guided by the Burkean concern that change must meet the overall test of whether it contributes to the long-term well being of the community. And the idea of the "well being of the community" means taking responsibility for renewing the genuine contributions of previous generations.

One of the misconceptions underlying the constructivist approach to the classroom, and to Freire's approach to consciousness raising in community settings, is that critical reflection will lead to a consensus on the nature of changes that must be undertaken. Both Dewey and Freire continually expressed their commitment to democracy and dialogue, but at the same time viewed as reactionary anyone who made a case for the importance of tradition—even those that are the basis of the community's patterns of mutual aid and self-sufficiency. Dewey confronted this problem in *Liberalism and Social Action* (1935), and even though he was challenging the tradition of capitalism (which is very different from the community-enhancing traditions I am referring to) he was unable to recommend how to deal with capitalists who possessed a different way of understanding intelligence. Freire's solution for dealing with those who opposed the social transformations that he envisioned as the outcome of critical reflection was to win them over through dialogue—which has not had a particularly good record of success. Neither Dewey nor Freire offer an adequate way of taking account of the possibility that the person or group who resists such changes as the widespread adoption of computer-based learning may have a clearer understanding of what form of education contributes to the long-term well being of the community. This is a particularly good example for bringing out the inadequacy of the different constructivist interpretations of the connection between learning and democratic decision making. The constructivist classroom, as discussed earlier, emphasizes the students as the primary decision makers, since their decisions are based on their own insights and understanding of relationships. Given that most students spend many hours a day using the computer to communicate with friends, surf the internet, write papers, download papers, and acquire information, they would likely support the even wider adoption of computer-mediated education. There is nothing in the constructivist textbooks used in the teachers' professional courses that suggests the importance of students learning to take account of the knowledge of the older members of the community, or to learn about the traditions that the older generations want to retain—which are sometimes in opposition to the industrial approach to agriculture, forests, food preparation, health care, and so on.

While the voices of students are too often ignored in political decision making at all levels of government, the ideological orientation underlying constructivist learning theorists would lead to yet another set of false and unfulfilled expectations: namely, that because students supposedly construct their own knowledge and values they should not have to compromise their autonomous decision making in order to accommodate what they too often consider to be the ideas of older members of the community who have not been taught to think for themselves and who are afraid of change. Dewey's method of intelligence accommodates the insights of all members of the community as long as they subscribe to the use and outcome of the scientific method of inquiry. But the other constructivist theories, as discussed earlier, reject the idea that knowledge can be transferred or shared. And their references to "renaming," "becoming," "open systems," and "autonomy" further strengthen the idea that anyone who questions the impact of innovations on the traditions of the community is likely to be viewed as a reactionary person and dismissed out of hand.

This emphasis on the student's construction of knowledge puts out of focus the need to learn about the traditions that students have been socialized from birth to take for granted. There is a further consequence of ignoring the traditions that make up the woof and warp of community relationships and activities; namely, by ignoring the traditions that students take for granted, the teacher is unable to help them recognize the limits of their own knowledge. The destructive traditions are also more likely to go unnoticed, which involves an irony not noticed by constructivist learning theorists. That is, while teachers are reinforcing students for constructing their own ideas, the students will be re-enacting many of the destructive taken-for-granted traditions of their culture. We have only to look back at recent history when students were encouraged to think critically, but in the process never became aware of racist and sexist attitudes and practices until political movements outside of the educational institutions forced teachers and professors to be aware of them. Today, we need to ask how many students, in constructing knowledge based on their own interests and unrecognized taken-for-granted cultural assumptions and practices, would recognize how the civil institutions that are now under threat represent sources of resistance to the rise of fascism. Indeed, the question needs to be asked if the students' own construction of knowledge would lead them to an awareness of the characteristics of fascism—and to how it can (and has) emerge through a democratic process.

The fact that critical reflection and individually constructed knowledge (which is more myth than reality as I will explain shortly) does not always lead to awareness of taken-for-granted cultural patterns can be seen in the silences of the leading constructivist theorists. All of them reproduced the tradition of thinking of cultures as evolving from primitive to advanced stages of development where decisions are based on some form of critical thought. And they all ignored the cultural non-neutrality of technology—which remains today a major silence in the writings of their followers and in constructivist classrooms. Other examples of cultural traditions they not only took for granted, but relied upon on a daily basis, can easily be cited.

⊠ Role of Language in the Cultural Construction of "Reality"

There is another aspect of culture that cannot be separated from the nature of traditions and from why it is so difficult to recognize the traditions we take for granted. This aspect of culture also involves a tension between the culturally prescribed patterns and the individualizing of these patterns in ways that reflect the person's self-identity and integrity—and what just seems to be the intuitively right thing to do. Recognizing this tension, which avoids representing culture in terms of genetic or linguistic determinism, can be understood most readily in terms of an insight attributed to Martin Heidegger that "language thinks us as we think within the language." To put this another way, we think within the conceptual possibilities of the categories and system of causality that are made available in the language of our culture. Sometimes thinking involves reliance upon what has been maginalized in the dominant culture or other cultural ways of thinking. Even widely misunderstood experiences can lead to changes in the language and thus in thought and behavior. This view of language represents a major challenge to the various constructivists' theories of learning, to the classroom teacher's romantic and ideologically grounded idea of student autonomy, and to the university professor who holds students accountable for expressing their *own* ideas. A case could also be made that this view of language brings into question the idea that people originate ideas and thus own them as private property. Again, in explaining the constitutive role of language I will focus on the characteristics that are particularly relevant to understanding the teacher's role as an intercultural and intergenerational mediator. This will also help clarify the other misconceptions and silences that are shared by constructivist-learning theorists and classroom teachers.

Constructivist theorists have based key aspects of their theories on a misconception about the nature of language that can be traced back to John Locke's idea that we put our personal meanings into words and then send them to other people. This sender/receiver model of the language and communication process, which Michael Reddy referred to as the "conduit" view of language, has been widely accepted in the West. Indeed, this view of language is at the center of how we think about computer-mediated communication; and it is also vital to maintaining other myths that still hold sway within the academic community—such as the idea that rational thought is free of cultural influence as well as the idea of objective knowledge. The tradition (in this case, a highly problematic tradition) of thinking of language as a conduit is especially important to maintaining the keystone of the constructivist position that represents intelligence as an attribute of the individual—or, in Dewey's case, as individuals sharing their intelligence in a way that conforms to the scientific method of inquiry. The conduit view of language also helps to deflect attention from the way in which constructivist-learning theorists share so many cultural patterns of thinking that have been carried forward over hundreds of years, and are now being imposed on nonWestern cultures by constructivist-oriented teachers.

We can most easily recognize how language thinks us as we think within the language of our cultural group by considering the ways in which language encodes and thus reproduces the metaphorically constructed thought patterns of previous generations. Understanding the dynamics of this process should be, as I have argued for years, a core part of the teacher's professional knowledge, particularly when we recognize that the teacher is mediating between different cultural ways of knowing and between the different generations that constitute the community of which the students are members. The basic dynamic that needs to be understood is how the root metaphors of different cultures provide the taken-for-granted interpretative framework that influences the process of analogic thinking and how over time the analog that is viewed as providing a more satisfactory way of understanding new experiences or phenomena is encoded in simplified form in the image words that become part of daily thought and communication. And as teachers understand this process they will then recognize that intelligence is cultural in the sense that the student's thinking is largely a matter of giving an individualized interpretation (and often misinterpretation) that is influenced by the conceptual categories and assumptions that have been passed on as part on as the taken-for-granted way of thinking. A perhaps even more important insight is that intelligence is a combination of the linguistically encoded intelligence and ignorance of previous generations, as well as the student's ability to understand relationships, have insights, question, and arrive at a sense of meaning. That previous expressions of a culturally specific form of intelligence is encoded in the language that influences thought at a taken-for-granted level of awareness can be seen in the many instances where the student's intellectual potential has been undermined by the earlier forms of intelligence (way of thinking). There are many examples of this process, such as when the language handed down from the past represented women as intellectually incapable of being artists, historians, scientists, and so forth. Another example, as we shall see later, is the historically encoded intelligence that represents all forms of life as mechanistic in nature, which we find current leading edge scientists basing their research and extrapolations on. To make the point in a way that sounds more familiar: the language in the textbook and computer software can contribute to reducing the student's potential intelligence to the level of past misconceptions and prejudices.

In the nineteenth century, Friedrich Nietzsche observed that we cannot understand something new (an experience, technological innovation, newly encountered phenomenon) on its own terms. Rather, the understanding of what is new is influenced by the interpretative framework the observer brings to it. Put another way, the new is understood in terms of the already familiar. This process of thinking in terms of similarities and using words that highlight them, is called metaphorical thinking. It is an imaginative process that fits the new into the already taken-for-granted schema of understanding. That is, metaphorical thinking is an "as if" way of thinking. Examples of how the already familiar involves using image words that help us understand new phenomena can be seen in the sciences where "black hole," "big bang," "cosmic dust," and so forth are used to convey an understanding of the phenomenon's basic characteristics. Other examples include thinking of the way in which computers process data as "artificial intelligence," the manipulation of the genetic characteristics of plants as "genetic engineering," and seeds that have been genetically altered in ways that make them sterile (thus forcing farmers to buy new seeds each year) as based on "terminator" technology. Previous experience and ways of thinking become the basis of understanding that a machine can exercise intelligence, that scientists can engineer new forms of life, and that life processes can be terminated. When the image words (iconic metaphors) are associated with the new phenomena, we then have a shorthand way of understanding that what is unique is similar to what we already understand.

If we pay attention to the spoken or written word among speakers and writers of English we find that sentences, as Vygotsky pointed out years ago, involve the stringing together of image words—words that carry forward the core concept or image worked out in an earlier process of analogic thinking. An example of how image words are strung together in a way that reproduces a particular way of understanding (one that reflects the assumptions and economic interests of a particular group) can be seen in the following textbook explanation that most students will be reading for the first time: "Silvaculture means *growing crops* of *trees*. It is *farm-*

ing in the *forest* instead of *farming* in the *field*." The image words (iconic metaphors), which have been put in italics, represent silvaculture "as like" a form of farming. The two sentences taken together represent the process of analogic thinking—that is, understanding the new in terms of the already familiar. Like so much of the analogic thinking that is part of teaching and learning, this analogy is highly problematic for a number of reasons that most students are not likely to recognize if this is the first explanation of silvaculture they receive.

An example from a book explaining the future of capitalism shows how analogic thinking relies upon the use of image metaphors that serve, even for the more sophisticated readers, as a basis of understanding—and misunderstanding when the analogy is more dissimilar than similar. Observe how Lester Thurow, as professor of economics at MIT, relies upon analogic thinking to explain one aspect of capitalism: "When technology and ideology don't smoothly mix, economic magna fluxes. Tectonic plates are violently thrust into each other—volcanoes erupt, earthquakes shatter the earth's crust, mountains rise, valleys fall. . . . The economic surface of the earth, the distribution of income and wealth, is now fundamentally remade" (1996, p. 18). Thurow's selection of an analogy, borrowed from the scientific understanding of the movement of tectonic plates, will lead many readers to a basic misunderstanding: namely, that just as humans have no control over the movements of the tectonic plates, they have no control over the economic policies that influence the distribution of wealth. Students and more general readers who are learning how to understand the nature of economic change are likely to accept the comparison between geological and economic processes as valid. That is, in the initial process of socialization, thinking is highly dependent upon the language and the schemata encoded in the metaphors that are made available.

Another point needs to be made about how the use of image words influences the individual's understanding. That is, image words such as "individual," "tradition," "data," "intelligence," and so forth, have a history—and when used in the present they reproduce the way of understanding from an earlier period in the culture's history. And the schema of understanding that has come down to us over time is the one that prevailed over other competing analogies—which always involves different economic and political forces, as well as the influence of earlier ways of thinking encoded in the metaphorical thinking of that time. This process of competing analogies and the social forces behind them can be seen in how the nature of a person's intelligence was to be measured. An "objective" measure of a person's intelligence was supposed to be more democratic and egalitarian than the old system of using social class and family connections as the basis for deciding a person's educational and work prospects. Measuring the size of the person's skull, which was based on analogic thinking that associated size with degree of intelligence, was abandoned when it was discovered that Western people did not possess the largest brain cavity. Eventually it was agreed that performance in an English language test would provide the objective, scientifically determined measure of a person's intelligence. The analogy where measurable performance results were taken as evidence of intellectual ability became simplified and encoded in the iconic metaphor of "intelligence score," which teachers and other professionals used for decades in channeling students into educational pathways that influenced their prospects for life. Now many psychologists are challenging this earlier process of analogic thinking that led to measuring intelligence by suggesting that the use of new discoveries in genetics provides a more scientific basis for understanding the nature of intelligence. Thinking of genes as determining intelligence leads them now to predict that a certain percentage of a person's intelligence is genetically determined. The root metaphor that drives this process of analogic thinking reintroduces the Social Darwinism of the late nineteenth and early twentieth century by providing a basis for determining which individual, cultural patterns (memes), and institutions are better adapted and thus will survive while other less well-adapted individuals and cultures disappear.

The way in which image words are associated with different previously taken-for-granted ways of understanding and personal experiences can be seen in how image words, even in the English language, are associated with different analogies. For example, Britains and Americans use different image words for referring to different parts of the car—and many other aspects of daily life. When we take into account the differences between cultures in the use of root metaphors that influence the layered nature of metaphorical thinking, from the selection of analogies to the encoding characteristics of iconic metaphors, we can see that the metaphorical characteristics of language can be a source of misunderstanding, such as when the Japanese student visiting in Louisiana was shot because he did not understand the meaning of the iconic (image) metaphor "freeze." The hidden process of analogic thinking encoded in image metaphors can also be the source of cultural imperialism. Image words taken for granted by Western educators and politicians, such as "emancipation," "progress," "modernization," "development," "individualism," "transformative learning," "globalization," and so forth, encode culturally specific assumptions that are carried forward and are now being introduced into other cultures as a more "enlightened" and "rational" way of thinking.

The aspect of metaphorical thinking that is most difficult to recognize because it provides the meta-cognitive schemata that influences the choice of analogies, and thus the meaning of image words, are the root metaphors. The metaphorical image of "root" is used to suggest that they are the source of a particular pattern of thinking that influences many, if not all—depending upon the culture—aspects of daily life. The image of a root also suggests that it is deeply rooted in the culture's history—in many instances, going back to the mythopoetic narratives and powerful evocative experiences that serve as the culture's earliest model of understanding. A culture's root metaphors (some cultures may be based on only one root metaphor) are the basis of its knowledge system and moral codes. They also influence the culture's way of thinking about space, design, patterns of social interaction, and, as we shall see later, even how scientific research is understood. To reiterate a key characteristic of root metaphors, their influence as a cognitive and moral schema is seldom noticed: that is, they are the bases of our natural attitude.

The two most powerful root metaphors in the West can be traced back to the book of Genesis. Patriarchy and anthropocentrism, which are represented as God's prescriptions for understanding male/female and human/nature relationships, undoubtedly had their origins much earlier in the oral narratives. The mainstream of Western culture has been deeply influenced by other root metaphors that have guided thinking about education, medicine, architecture, agriculture, business, and so on. These root metaphors include the idea that the individual is the basic social agent and thus the source of rational thought and moral judgment—and personally accountable for economic success.

Equally powerful as a taken-for-granted way of thinking is the root metaphor that represents change as inherently progressive. Economism—that is, thinking of relationships, activities, and the environment in terms of their monetary value and how ownership helps one get ahead—is becoming increasingly dominant as a root metaphor. And just as some root metaphors are being challenged, such as patriarchy and anthropocentrism, other root metaphors are gaining wider acceptance as a schema for understanding biological and cultural processes. Two contending root metaphors that are gaining ground among different segments of the population are evolution and ecology. As many interpreters of evolution think of it as the expression of a linear form of progress, it has profoundly different political and economic implications from how others are basing their understanding of culture on the root metaphor of ecology.

The influence of each of the above root metaphors can be seen in the analogies selected for understanding new phenomena, for solving new problems, and in traditions that range from legal procedures to the design of the material culture. But it is in the spoken and written language that we can most easily see the influence of root metaphors on patterns of thinking and moral values. Nietzsche's description of metaphorical thinking as "fitting the new into old schemas" can be seen in how the root metaphor of mechanism underlies the thinking of highly intelligent thinkers working at the cutting edge of their fields of endeavor. The following examples were selected because they demonstrate how a root metaphor provides the basic interpretative framework that is reproduced over hundreds, even thousands of years in terms of some root metaphors. They were also chosen because they serve as examples of how even the most rational and highly educated scientists unconsciously reproduce a culturally specific pattern of thinking that introduces fundamental misconceptions into scientific explanations.

> "My aim is to show the celestial machine is to be likened not to a divine organism but to a clockwork." Johannes Kepler (1571–1630)

> "Our conscious thoughts use signal-signs to steer the engines in our minds, controlling countless processes of which we're never much aware." Marvin Minsky, 1985

> "But another general characteristic that successful genes will have is a tendency to postpone the death of their survival machines at least until after reproduction . . . Survival machines began as passive receptacles for the genes, providing little more than walls to protect them from the chemical warfare of their rivals and the ravages of accidental molecular bombardment." Richard Dawkins, 1976

> "The machine the biologists have opened up is a creation of riveting beauty. At its heart are the nucleic acid codes, which in a typical vertebrate animal may comprise 50,000 to 100,000 genes." E. O. Wilson, 1998

It would be just as easy to quote other individuals who represent a similar span of time. For example, Thomas Hobbes, the author of *Leviathan* (1651) and one of the "fathers" of modern liberalism wrote "For what is the heart, but a spring; and the nerves, but so many strings; and the joints, but so many wheels, giving motion to the whole body." In using the root metaphor of mechanism as a way of explaining the similarity between computers and human thought, Anita Woolfolk, writing in *Educational Psychology* (1993) states that "like the computer the human mind takes in information, performs operations on it to change its form and content, stores information, retrieves it when needed, and generates responses to it." Hundreds of years separate Hobbes and Woolfolk, yet they both rely upon the same root metaphor to explain different phenomena, just as Dawkins and Wilson repeat the same deep pattern of thinking that Issac Newton and Kepler introduced as a way of overturning the root metaphors that were the basis of thinking and social life in the Middle Ages. These examples can also be understood as traditions that are carried forward by individuals who find them useful—even though they are fundamentally problematic in ways that these leading thinkers do not recognize. They are also examples of how language thinks us even as we think within the language—to recall an earlier statement that may seem vague without concrete examples.

Individuals may extend the explanatory power of a root metaphor to new areas of cultural experience, which can be seen in how the Bauhaus school of architecture derived their principles of design from how they understood the characteristics of the machine. The machine metaphor (industrial model) has also been extended to agriculture, medicine, education, forestry, and, now, to the creation of new forms of life. The way in which a root metaphor frames what will be given attention and what will be ignored can be seen in the proposal of Lee Silver, a molecular biologist at Princeton University, who explains how genetic engineering can be used to create a separate class of "Gene-Rich" people who will become the governing class, while the underclass, whom he refers to as "Naturals," will do menial work. Engineering (which is based on a mechanistic root metaphor) puts out of focus the moral and political issues that are not usually associated with machines. But the root metaphor of progress, which gives legitimacy to the many uses of the mechanistic root metaphor, is clearly present in the thinking of Silver, and the other scientists quoted above. The idea of progress and the idea of building better machines or machine-like processes are mutually supportive, which leads to ignoring whether the machine-driven form of progress is a major contributor to the many environmental problems we are now facing.

Root metaphors have explanatory power, especially when they totally dominate a way of thinking. And they are often the basis of power relationships that benefit one segment of society over others. Patriarchy and evolution are two powerful examples of the political role that metaphorical thinking plays. If we consider the root metaphors (mythopoetic narratives) of more ecologically centered cultures, we find that they often serve as the basis of moral reciprocity within the community and with the non-human world. And there are root metaphors in other cultures that explain life as a cycle within larger cycles, which influences how birth and death are understood—as well as other cultural patterns. Root metaphors, to sum up, are specific to different cultures. But some root metaphors, particularly those that underlie the high-status knowledge that drives the current phase of globalizing an industrial approach to production and consumption, undermine other cultural ways of knowing, their capacity to be self-reliant, and the environment they depend upon.

One of the implications for educators is that the examination of root metaphors needs to become part of their professional responsibility. That is, they need to recognize in the curriculum how earlier patterns of thinking influence the thought processes of textbook writers and the people who write educational software. Most importantly, they need to recognize the root metaphors that are an inescapable basis of thinking. The following summary represents some aspects of root metaphors to which classroom teachers and professors need to give special attention.

(1) Root metaphors, as the above example of mechanistic thinking demonstrates, influence what will be given attention, and what will be ignored. For example, when the mechanistic root metaphor is used to explain mental processes, it marginalizes the cultural influences on conscious-

ness, intentionality, self-identity, feelings, meaning, and value judgments. When it is applied to agriculture it marginalizes the interactive and interdependent nature of the life forms that make up the local ecology—and leads to technological interventions that disrupt the complex information exchanges vital to the interdependent life processes in healthy ecosystems. The root metaphor of patriarchy, to cite another example, made it difficult for people to recognize that women could be artists, historians, scientists, skilled craftspeople, and so on. The root metaphor of evolution, when combined with the root metaphor of linear progress that we find in the thinking of some scientists, represents the forces contributing to globalization as expressions of nature's process of design. At the same time the ecologically centered cultures that are essential to sustaining biodiversity are represented as poorly adapted and on the way to extinction. When thinking is controlled by the root metaphor, the silences that should be examined are generally ignored. We can see in this process both the influence of tradition and the taken-for-granted nature of most of our thinking patterns—even on the thinking patterns of those who supposedly originate their *own* ideas.

2. Root metaphors are the basis of a culture's moral values, which we will see more clearly when we consider the knowledge being lost in a constructivist approach to the education of Quechua children. We can also recognize how root metaphors influence the moral norms largely taken for granted in our own culture by considering the influence of root metaphors such as mechanism, anthropocentrism, patriarchy, evolution, and so forth. The question that needs to be asked in making explicit the influence of our dominant root metaphors is "How does the root metaphor represent the attributes of the participants in the relationships that make up everyday life?" When the root metaphor of patriarchy represents the attributes of women as inferior to those of men, it sanctions the subjugation of women as normal—that is, as a taken-for-granted moral norm that gets passed along through the languaging processes over many generations. When anthropocentrism is the guiding root metaphor for understanding the relationship between humans and the natural world and the attributes of the participants in this relationship, it sanctions as moral the exploitation of the environment. And the root metaphor of progress sanctions as moral many practices that in other cultures, especially ecologically centered cultures, would be regarded as immoral. Again, the critical questions to ask about root metaphors have to do with what is put in focus, what is taken for granted, and what are the silences. Other questions include asking about how relationships and attributes of participants in the relation-

ships are understood, and who gains and who is oppressed by the "reality" and moral norms constituted by the root metaphor.

In the chapter on the teacher's role as an intercultural and intergenerational mediator we shall return to a consideration of how root metaphors influence every aspect of the curriculum and the responsibilities of teachers that go well beyond the facilitator role that constructivist theorists assign to them. For now, given the above discussion of the constituting role of root metaphors, it is important to return to a further consideration of the misconceptions of constructivist learning theorists.

One of the tests of a theory of learning is that it should explain the thought processes of the theorists themselves. That is, their own way of thinking should not contradict their theory. However, when we consider whether Dewey, Freire, Piaget, and their followers constructed their own critically based knowledge we find that their deep, taken-for-granted patterns of thinking were based on the same root metaphors that have given conceptual direction and moral legitimacy to the industrial-based culture that is now being spread with messianic fervor to even the remotest cultures. The root metaphor of evolution, which was and is again being used to explain cultural developments, is the basis of how Dewey, Freire, and Piaget explain different levels of intelligence. Each of them represented their own approach to the construction of knowledge as the more evolved form of intelligence. It would not be inaccurate to claim that their evolutionary pattern of thinking has contributed to the mainstream, Western culturally held belief that nothing can be learned from cultures that have not reached the West's stage of development. And a further generalization is warranted: namely, that this bias against learning from nonWestern cultures is repeated in the way the promoters of constructivist-based educational reforms in nonWestern counties ignore the importance of including local intergenerational knowledge in the curriculum.

We also find that Dewey, Freire, and Piaget interpreted change as the expression of a linear form of progress. Furthermore, they shared the cultural assumption that represents the individual as the basic agent of critical reflection and constructor of new knowledge. Dewey had a somewhat different interpretation that represented knowledge as a social construction, but he shared with Freire the assumption that knowledge must be newly discovered and not acquired from the past. The combined influence of a taken-for-granted way of thinking of the never-ending process of reconstructing experience as a manifestation of progress led Dewey to dismiss tradition by saying "Let the dead bury their dead," and Freire to claim that "history has no power." Yet both were reenacting a tradition of thinking that was inconsistent with their theory. And the cultural influences on Piaget, given that he must have considered himself as having reached the stage of biological and cognitive development that would enable him to be an autonomous thinker, are equally obvious—if we care to look for them. To take another example, all three theorists were influenced by the root metaphor of anthropocentrism—which is now being repeated by the followers of Freire and Piaget. The current interpreters of Dewey are trying to make the case that he did not have this bias, but they ignore that his naturalized interpretation of intelligence did not lead to any suggestions that we can acquire any moral insights from the interdependent characteristics of natural systems—and he did not warn, as noted earlier, against trashing the environment that was well advanced by his time. Lastly, it should be noted that all three theorists were influenced by the root metaphor of patriarchy and thus used the gender-biased language of their era—which Freire apologized for late in his life.

The traditions of metaphorical thinking passed on in the writings of these Western "fathers" of the different interpretations of constructivist learning is strong evidence that their own patterns of thinking do not fit what they prescribed for others. And the same criticism can be made of the constructivist-oriented professors of education who have learned from the writings of Dewey, Freire, Piaget, and Whitehead to think in the same taken-for-granted patterns—even in an era where third world voices are becoming more self-confident and challenging Western assumptions, and where there are daily reminders of a deepening ecological crisis that is moving beyond the power of scientists to manage or reverse.

If the leading theorists and current promoters of constructivist-based educational reforms are unable to recognize the culturally specific assumptions their theories of learning are based upon, or to recognize the silences that are becoming by the day more significant, how can students be expected to recognize them? And how can students be expected to construct their own knowledge and thus emancipate themselves so that they can become autonomous individuals when they are exposed to a constant barrage of media messages scientifically engineered to influence the deepest levels of their consciousness and self-identity? The tragedy is simply being compounded by encouraging students to think they are constructing their own ideas, meanings, and identity, when this rootless form of individualism is exactly what serves the interests of the promoters of consumerism. Students may learn to think critically about aspects of their world, but they are not likely to understand that their own subjectively limited knowledge and lack of skills will not provide a real basis for resisting the forces of consumerism and environmental destruction. Like so much that is learned in Western public schools and universities, the abstract talk, critical reflection, and search for the latest intellectual fad will simply provide a ritualized response to the multiple crises we now face. And when this Western approach to education is adopted by or imposed upon non-Western countries, the possibilities of resisting the Western model of development will be further undermined.

We now need to turn to a consideration of the role of the teacher in different approaches to revitalizing the commons, in mediating between different cultural ways of knowing, in helping students sort out the gains and losses connected with adopting a Western form of modernism, and in acquiring a more complex and balanced understanding of intergenerational knowledge. We also need to clarify further the connections between constructivist educational reforms, the further destruction of the commons, and the globalization of a consumer-dependent lifestyle that few of the world's population have the resources to participate in and which further exacerbates the ecological crisis.

 chapter 4

How Constructivism Undermines the Commons

Given the previous explanation of how the languaging processes of the culture influence the student's tacit and explicit knowledge, and thus communicative competence, it is necessary to ask: What is destructive about the form of education promoted by the followers of the different constructivist-learning theorists? To put the question differently, Why would anyone question the optimism and faith that constructivist theorists have in the student's ability to learn in a way that enhances their individual autonomy? With the exception of Dewey, all of the constructivist theorists view youth as the source of new ideas and values—which can only be achieved by protecting them from the "closed-knowledge systems," "banking approach," and "spectator knowledge" of adults who supposedly have been victimized by earlier generations of adults.

Freire goes even further than the constructivists who adhere to the mistaken idea that "knowledge cannot be transferred" by making a universal claim that most current proponents of constructivism view as the expression of profound wisdom. As I think Freire's claim is both naïve and destructive of how the intergenerational knowledge that sustains the commons is renewed, I shall quote him in full:

> In my view it's preferable to emphasize the children's freedom to decide, even if they run the risk of making a mistake, than to simply follow the decision of the parents. . . . One of the pedagogical tasks of parents is to make it clear to their children that parental partici-

> pation in the decision-making process is not an intrusion but a duty, so long as the parents have no intention of deciding on behalf of their children. The participation of the parents is most opportune in helping the children analyze the possible consequences of the decisions that is taken. 1998, p. 97

Freire does not qualify this statement to take account of situations where the child's decision ignores the potentially harmful consequences that parents may be more aware of. Nor does he recognize cultural differences in the patterns of parental guidance.

Granted, there are authoritarian parents and cultural norms that severely limit the child's behavior and life chances. In questioning the context-free formula of Freire and the other constructivists that is supposed to lead to what Oliver celebrates as the "becoming of experience" (2002, p. 100), I am not giving my support to any of the many expressions of authoritarian relationships that exist between parent and child and between teachers and students. Nor am I in agreement with the way many parents fulfill a personal psychological need by living through the achievements of their children—which they often attempt to control for their own ends. The nature of authoritarian relationships, and how they differ from patterns of parenting in different cultures that contribute to the psychological well-being and communicative competence of children, is exceedingly complex. But this complexity is ignored by the dichotomous categories that characterize the thinking of the constructivists. Their prescriptions simply lead to another form of extremism characterized by reductionist thinking and simplistic guidelines that few parents, regardless of culture, would take seriously.

What I want to consider are a different set of consequences that result from the emphasis on students' constructing their own knowledge. That is, I want to consider how constuctivist approaches to education, which are now being promoted in English- and nonEnglish-speaking countries, undermine the commons. My concern leads to two more questions: What is meant by the commons? and How does the destruction of the commons contribute to the spread of the industrial approach to production and consumption that is undermining the viability of the earth's natural systems?

The idea of the commons can be traced back to a body of common law that existed before the Roman conquest in A.D. 43 of what was then called Britannia. The early understanding of the commons was that land, and its many resources, could not be divided and privately owned. Originally, the commons was the pasture land that could be used by all members of the community. But the idea of the commons encompasses much more and still exists in many cultures that can trace their understanding of the commons back to the earliest stages of human existence. When we identify what all the commons encompasses in addition to the land, we can see that it is integral both to healthy rural and urban environments—and to morally coherent communities.

By interpreting the commons to mean what is commonly shared between humans, and between humans and the non-human world, we can see that it includes the quality of the air, water, plant and animal life—in effect, all that is essential to a self-renewing ecosystem. In addition, the commons includes the symbolic systems that are shared in common and that are essential to the ability of different human communities to sustain and renew themselves. The symbolic systems include a wide variety of technological systems, spoken and written language, narratives that are the basis of the community's moral codes and the self-identity of its members, and the knowledge and aesthetic sensitivities that influence the community's approaches to food, music, and the other arts, ceremonies, and leisure activities. The commons, in effect, encompass everything that is not privately owned and that has not been turned into a commodity. Another characteristic is that what is shared in common is intergenerationally renewed—and modified. Keeping in mind the tendency of some readers who will find it easier to accuse me of romanticizing the commons than to examine the non-privatized and non-monetized commons they rely upon, I want to state unequivocally that some cultures are based on symbolic systems that are cruel, oppressive, and unjust. The symbolic systems of other cultures contribute to commons that provide for the basic needs of food, shelter, mutual support systems, and conceptual/moral frameworks that give its members a sense of belonging and purpose. To reiterate a basic point, the destruction of the commons–through the use of inappropriate technologies, belief systems that bring more of the commons under private ownership, and approaches to education that undermine intergenerational knowledge—diminishes the life chances both of the human and non-human members.

So far I have used rather broad categories to explain what all the commons encompasses. In order to recognize the destructive impact that the constructivist ideal has on the commons, I will focus on the commons that sustain the Quechua of the Peruvian Andes and the Balinese. I will also focus on the commons of rural and urban America. In narrowing the examples to these cultures, it needs to be kept in mind that a similar analysis could be done on how constructivist-based educational reforms undermine the commons in countries as varied as Turkey, Taiwan, Ukraine, and Brazil. Another observation also needs to be kept in mind: namely, that the analysis is not based on the constructivist assumptions that knowledge cannot be intergenerationally transferred, that the languaging processes do not influence the child's earliest patterns of thinking, that children do not learn by observing the behavioral patterns of peers and adults in their family and community. These constructivist assumptions ignore all the pathways of learning that constitute the interactive relationships that sustain the commons and that children interact with on a daily basis. Again, it needs to be stated that Vygotsky understood the social basis of learning, but his ideas are only given token treatment in constructivist textbooks—mostly for the purpose of giving further legitimacy to the constructivist dogma that the child's "reality" is individually constructed—and aided by peer participation.

The key point here is that since the child's everyday taken-for-granted reality is largely culturally constructed, the analysis here will focus on how constructivist pedagogies and curricula reinforce the Western bias against the intergenerational knowledge that is the basis of the commons of different cultures. The analysis will also be focusing on the destructive impact on the commons when students adopt the Western assumptions that are reinforced in constructivist classrooms. As the Western icons of a consumer lifestyle are being adopted by students in different parts of the world, we can expect that many of them will readily embrace the Western values and ways of thinking promoted in constructivist classrooms. Children and teenage youth in other cultures cannot be expected to resist the allure of Western images of material success and individual happiness when, for example, people rooted in the sophisticated Chinese civilization begin to build in the outskirts of Beijing gated communities with Western street names and California-style houses.

In discussing how constructivist based educational reforms undermine the intergenerational knowledge that is essential to the commons in third world cultures it is essential to address a deeply held bias by Western theorists—including professors of education. Vandana Shiva points out that this bias, which leads to viewing many nonWestern cultures as backward, is based on the assumption that the ability to participate in a money economy and to give evidence of years of schooling and other data relevant to a modern consumer lifestyle, is what separates modern from pre-modern stages of development. This leads, according to Shiva, to equating a subsistence lifestyle with poverty and backwardness. This bias overlooks important differences, such as the fact that locally grown grains are more nutritious than processed foods, that houses built with local materials are better adapted to local environmental conditions than those built with cement, and that natural fibers and local traditions of dress are more suited to local climates (1990, p. 97). Western developers and educational reforms, as she observes, too often confuse non-monetized local economies with the misery form of poverty that accompanies the disruption of the local commons by the introduction of Western technologies and integration into the Western industrial, market-oriented economy.

Another point needs to be made in order to ensure that another cultural bias is not imposed on the examples of how constructivist educational reforms are predicated on false promises that cannot possibly be fulfilled. This bias, most often expressed by university professors, other types of Western experts, and university students, is that any discussion of intergenerationally connected and non-monetized patterns of mutual aid in indigenous cultures represents romanticized thinking—and an escapist mentality that wants to go back to an earlier, less complicated stage of existence. Again, Vandana Shiva helped me to understand how to respond to the criticism. She pointed out that subsistence cultures are part of the living present and that they represent the majority of the earth's human population. An additional point needs to be clarified before considering the Quechua of the Peruvian Andes as an example of how constructivist-based educational reforms are part of the Western project of colonization. That is, while I am not suggesting that we should try to copy their cultural patterns, I do think we in the West can learn from them about the important connections between cultural diversity and biodiversity. Learning about their cultural patterns and achievements also helps us to see our own cultural patterns more clearly.

⬦ The Commons of Indigenous Cultures and Non-Western Cultures

The culturally disruptive and colonizing nature of constructivist approaches to learning can be seen in the Peruvian government's most recent attempt to bring about educational reforms. Basing reforms on the Piagetian approach to constructivism is the latest in a series of reform efforts that go back to the nineteen sixties. In this latest effort, teachers are being trained in the Western pattern of thinking required in subjects such as mathematics and physics and in the constructivist principles of learning—which they are expected to base their lessons on in rural and urban classrooms. In addition, teachers from urban areas who are being sent to schools in indigenous communities often lack knowledge of the Quechua and Aymara culture of the children. The combination of a curriculum that is largely based on traditional Western subjects areas (65 percent according to governmental regulations), the inability of teachers to communicate with students in their native language, and the use of teaching strategies based on the assumptions that students will construct their own knowledge from their own encounters with the environments organized by the teacher, could not be a better formula for failure—for all participants. Both students and teachers are caught in a double bind that is never resolved, but nevertheless has disruptive consequences for the community.

In order to avoid misinterpreting the Quechua and Aymara (the latter live mostly in the Lake Titicaca border area of southern Peru and northern Bolivia) as examples of indigenous cultures that have worked out a few things that Westerners find economically useful to patent, a few observations are in order about what they have accomplished in the mountains, valleys, and plateaus that range from sea level to over 15,000 feet in elevation. These cultures have been traced back over 8,000 years, and during this period they have developed an incredibly complex relationship with the commons. Their understanding of the varied characteristics of microclimates and ecological niches has led to the development of one of the world's most diverse approaches to the cultivation of plants. They cultivate over 3,000 varieties of potatoes and an equally impressive number of other tubers. European foundations now recognize the Quechua as critical to maintaining the biodiversity of the Andes and support Peruvian NGOs that are dedicated to helping the Quechua and Aymara renew their traditional knowledge—which is now being threatened by the pressures of modernization.

The Quechua and Aymara cultures are based on a profoundly different way of thinking than we find in the West. The differences can be seen, in part, by comparing how Quechua and Aymara children learn with the assumptions underlying constructivist principles of learning. One of the first differences that stands out is that Quechua and Aymara children learn to be communicatively competent within the cosmovision of their culture and to take on social responsibilities at an age that does not correlate with Piaget's stages of cognitive development. And the intergenerational basis of their learning does not correspond to the Deweyian/Freirean view that emancipation and the continual reconstruction of the traditions of previous generations represent the highest goal of education. In providing examples of what and how Quechua and Aymara children learn within the context of their culture's belief system, it is necessary for the reader to keep in mind that a fully adequate account of their cultures cannot be provided in a few pages. For readers who want a fuller description of indigenous cultures by Peruvians who have lived their whole lives among them and have studied and written about them, they should read the collection of essays in the book edited by Frederique Apffel-Marglin (with PRATEC) titled *The Spirit of Regeneration: Andean Culture Confronting Western Notions of Development* (1998).

Perhaps the best way to introduce the profound differences that separate the assumptions and strategies of constructivist-learning theorists (including Piaget, Dewey, and Freire) from the way of knowing that influences what and the way Quechua and Aymara children learn is to consider what Carlos Ortega Hores (a ten-year-old boy from the district of Plateria, Puno) says about his knowledge and responsibilities:

> My parents taught me how to nurture the animals: to graze and lead them to the places where the pastures grow. I know all the animals that my father nurtures. I know how to ride a horse . . . I know the diseases like diarrhea of the offspring which we know how to treat with creosol. Mother alpacas are covered at two years of age and at three they are already giving birth . . . We do *Uywa chuwa* (a welcoming ritual in the nurturing of alpacas) at

Christmas time. I know how to cook and sleep alone tending the animals. Rengifo, 2001, p. 12

Carlos did not learn about the practices essential to maintaining the health of the animals through critical or even experimental inquiry. Nor did he acquire this knowledge on the basis of his subjective experience. Rather he learns, like other Quechua and Aymara children, from watching parents and others in the community, from conversations that are part of participating in the work and festivals of the community, and from listening to the sounds and observing changes in what we call the environment—which the Quechua and Aymara consider to be persons like themselves who are engaging in an ongoing conversation. As a group of Smith College students report, based on their field research among the Quechua done under the guidance of anthropologist Frederique Apffel-Marglin, learning is largely embodied and mimetic. Just as the Quechua farmer does not rely upon a chemical analysis of the soil to know if it is "tired," but on feeling the soil with his hands, the child also learns about the condition of the soil in the same way. Mario Arevalo, a former educator and founder of the NGO, Pradera, further clarifies how Quechua children learn by noting that Quechua knowledge is not viewed as located in the head alone, but in the farmer's hands, eyes, nose, soul—and in the ongoing dialogue with the other forms of life that are members of the commons (Shmulsky, Marlowe, Daniel, 2003, p. 3).

Jorge Ishizawa Oba notes that one aspect of the dialogue with the community of nature (*sallqa*) and the community of deities (*huacas*) is learning to read the signs that are part of the mutually nurturing relationships. The form and moment of flowering, the place and the way birds nest, the behavior of vicunas and foxes, the mode of appearance of the constellations, as he notes, are part of the dialogue and mutual nurturing that tell the Quechua if the rains will be abundant or scarce. He gives a further example of how Quechua decisions are dependent upon the ongoing dialogue that includes reading the signs communicated by nature. In the altiplano of Puno, llama and alpaca herders observe how many eggs are layed by a small bird called the chijta. If the bird lays three eggs the herders know that there will be abundant forage, and thus they will take steps to ensure that the females in the herd have as many offspring as possible. If there is only one egg, the herders will reduce the size of the flock because they know there will not be enough forage (Ishizawa Oba, with Grillo Fernandez, 2002, p. 25). The dialogue is an ongoing part of daily life—and thus the Quechua are constantly learning how to nurture and to be nurtured.

In the Quechua community of Lamas, and in other communities throughout Peru, children begin to learn the Quechua way of being before birth. From conception on, the fetus is linked to nature through the plants the women eat and the nocturnal songs of male and female birds that announce whether a boy or girl will be born. Unlike the constructivist view of learning where "becoming," "emancipation," and the ongoing reconstruction of experience liberates the child from the knowledge of the older generation and where the stages of cognitive development are genetically dictated, the Quechua view children as born with an inherent wisdom that makes them receptive to learning the reciprocal relationships that exist between all members of the *ayllu*—the community of relatives made up of human persons, the members of nature, and the *huacas* or deities (Rengifo, 1998, p. 90). The child's understanding of reciprocity, nurturance, and dialogue begins as the baby, strapped to the back of the mother, watches her actions in the field. When the infant is too heavy to carry, it accompanies the mother or father in the planting, rituals, festivals, and harvesting. One boy told the Smith College researchers that at the age of three his grandfather taught him the names of plants, and at the age of five he began working in the *chacra* (the plot of land). By age twelve he had his own chacra and presented the agricultural products he had grown at the local Seed Fair. This twelve-year-old's level of knowledge and responsibility is not the exception among Quechua and Aymara children. Four-year-olds begin to learn how to spin wool, and by age six they possess the knowledge necessary for spending the day alone watching the cattle graze. At age eight or nine, girls become active in the preparation of food, and, as one Smith researcher reports, a ten-year-old girl was instructing others in ceramics.

Constructivist learning theorists are likely to view Quechua and Aymara children as being exploited and oppressed by ancient traditions that Dewey associated with the spectator approach to knowledge and that Doll refers to as the intellectually stultifying "closed system." To most Western observers, what the Quechua and Aymara child learns may appear as tradition-based and as static. But this perception could not be more wrong. The child's way of learning is embedded in the intergenerational knowledge that has been tested, refined, and accumulated over thousands of years. But it is not the spectator relationship with a fixed body of knowledge as Dewey suggests. And it does not fit the image of transferring knowledge that Freire called the "banking approach" to education.

To understand the continuous nature of learning that begins in the earliest years of childhood and continues through life, it is necessary to understand several key aspects of the Quechua and Aymara world view—or what close observers of their culture refer to as their "cosmovision." In order to avoid misrepresenting Quechua and Aymara, I will quote Grimaldo Rengifo, a lifelong student of Quechua culture and founder and co-director of PRATEC (Andean Project for Peasant Technologies): "In the Andean cosmovison, the world is perceived as a dwelling for a multiplicity of animated beings. The mountains, the waters of rivers, the rains, the *huacas* or deities, the plants, the animals, the human community, and the wind all have life." He goes on to identify another key difference between the Quechua and Western way of understanding. Given that the Quechua understand that everything is a relative to everything else and necessary to the re-creation of life, it then becomes necessary to pay constant attention to what all the living participants are communicating—and to the nurturing nature of the relationships. As Rengifo put it:

> In this way of living, each one is showing his or her way of being, each one is saying, is speaking. Each being expresses itself through 'signs' that are continually saying something about themselves and how their relations with other feel. The position and luminosity of the stars, for example, 'tell' the farmer about the aspects of the climate because the behavior of the stars in that moment corresponds to a mode of dialogue with other beings. Similarly the frequency, intensity, and colour of the winds are 'telling' the farmer of the weather that is about to happen, because the wind presents a colour and odour that express the particular way it dialogues with other beings. Nature 'speaks' to the farmer, just as the farmer 'speaks' to nature and the *huacas* about diverse matters. There is no objectification here, no distancing from nature on the part of the human community; rather, it is 'a dance' where all are dancing fraternally to the sound of a natural music, in a cosmic dance in which we participate and to which we contribute. 1998, pp. 176–177

Children learn to "read" the signs being communicated by every aspect of their living and dynamic environment. But it is not a matter of the passive, mentally centered relationship that characterizes reading in the West that relies upon individual interpretation. It is more of a constant dialogue, where the 'signs' are understood as expressions of nurturance—and call for reciprocity on the part of humans. As everything is understood as in a state of regenerative change, learning is continuous—but in the Quechua and Aymara way. Thus, traditions—festivals, rituals, ways of being open and respectful in interpreting the language of nature—provide guidance in how to be in a reciprocal and nurturing relationship with the rest of the community or what can be called the commons. Far from being static, these traditions have enabled both the Quechua and Aymara to live symbolically rich and ecologically sustainable lives for thousands of years—and to make an important contribution to expanding biodiversity in ways that have benefited the rest of the world's population.

In considering how constructivist educational reforms represent yet another attempt to colonize the Quechua and Aymara by socializing their children to think in terms of Western assumptions and materialistic values, it is necessary to identify yet another key difference—which is how they view the individual. Unlike the constructivists' ideal of the autonomous individual (and the generational autonomy envisioned by Dewey), the Quechua and Aymara do not think of the child as beginning the lifelong quest of attaining a greater degree of individual autonomy through critical inquiry. Nor do they view life as seeking a better footing on the endless treadmill of "becoming," as envisioned by the followers of Whitehead such as Doll and Oliver. None of the pedagogical strategies that constructivists represent as the one-best way of separating the students from their cultural traditions applies to Quechua children and their cosmovision. For them, embodied and respectful relationships with all living beings are the source of meaning, identity, and nurturance. Rengifo explains this relational view of the individual in the following way:

> There are no child rights which are not at the same time rights of her family, of her community. The child and any member of the *ayllu* (extended family including human relatives, nature, and deities) is perceived in terms of her relationships with the community and not with outside of it. It is with the community, her family, her *ayllu*, that life of each member has meaning. By making visible the values of children there emerge also the values of her community since in her the community lives. 2001, p. 13

That the cosmovision of the Quechua and Aymara is not based on a linear view of change and time, but represents change in terms of the natural cycles of the seasons, it is also important to understanding their view of life as continuous participation in the dynamic network of nurturing relationships. By way of contrast, the view of time taken for granted by all of the constructivist-learning theorists is that time is linear—and that it can be wasted in ways that undermine the individual's progress in attaining a higher state of autonomy and efficiency in reconstructing experience. With the exception of Dewey, who as we noted earlier, wanted all cultures to adopt his community-centered experimental method of inquiry, all of the other constructivists place the achievement of individual autonomy over the wellbeing of the community. The irony is that these apostles of the Enlightenment vision of emancipation were unable to recognize the parallel between Adam Smith's theory of how the "invisible hand" supposedly ensured that individual competition in the marketplace would uplift the entire community and their idea of how the ongoing process of emancipation would improve the quality of life in the community. And like Smith, their idea of what constitutes the basis of individual agency ignores the fate of the non-human members of the commons.

Given the profound differences between the indigenous Andean and the Western approach to knowledge being prescribed by the Peruvian government, it should be no surprise that in most indigenous communities there is an uneasy relationships between teachers and parents. While many parents see the necessity of schools, and some even view schooling as providing their children the basis of earning a living in a Western-style job, there is also concern that the school will alienate their children from the traditions of the family and the community. This uneasiness about the alienating effects of government-sponsored schooling can be seen the in the range of criticism that teachers direct toward the community and that parents and other community members direct toward the teachers. When teachers become more educated in the use of constructivist approaches to teaching both the Western subject-oriented curriculum and the local knowledge component, the criticism will become more intense as the future of Quechua and Aymara cultures will be more at stake.

For now, the tensions between the community and the school are rooted in issues of relevance and competence. For example, in a community in the northeast region of Peru, the teacher complains that too often when children go to the fields with their fathers they do not return to the classroom. The teacher also complains that the parents do not provide the school supplies needed by the students, and that parents cannot be counted upon to make repairs in the building and the fences. Parents, in turn, complain that their contributions to maintaining the school building too often go unrecognized, that the teacher is an outsider who does not understand Quechua culture, and that the teacher is too often indifferent to whether the students learn. Other parents are more supportive and worry that the community may lose their school (Costilla, 2003). But as students begin to exhibit the Western mind-set reinforced by constructivist approaches to knowledge, the criticism will become more focused on issues of cultural domination that the Quechua and Aymara have had to deal with over the last 500 years. Given the constructivist emphasis on forms of inquiry that are the basis of the Western industrial-based culture, and its built-in bias against all forms of intergenerational knowledge, some students may find it difficult to understand that the Quechua and Aymara traditions are more ecologically sustainable than the seemingly more dynamic and materialistic, wealth-creating Western culture. That is, they may adopt the Western idea that success is to be understood as the pursuit of individual wealth, rather than understanding that success and non-material wealth are achieved through participating in the commons where there are systems of mutual support and nurturing between all forms of life.

What constructivist theorists ignore is that there are still close to 6,000 spoken languages today, with many of them on the verge of extinction. Each of these cultural ways of knowing involve understanding the commons in different ways. The use of technology and patterns of self-sufficiency also differ widely—but many are notable for their smaller ecological footprint. For example, the Balinese religion involves a system of temples that play a key role in the cultural emphasis on the arts, which are essential to their festivals and ceremonies. The festivals connected with these temples also serve to regulate the allocation of water to the rice paddies, ensuring the distribution of water throughout the bioregion. The arts—puppet theater, dance, poetry, music, paintings, etc.—connect the present with the past and future in a way that keeps in balance the competing forces of growth and decay. In effect, traditional Balinese culture continues to represent a model of the commons that has not yet been entirely overwhelmed by Western-style individualism and consumerism. But this ecologically balanced culture is now being threatened by tourism and Western technologies that have been misrepresented as culturally neutral. Unlike Bali where constructivism has not yet been introduced, the government of Taiwan adopted constructivism eight years ago as the basis of the country's educational reform. But the Taiwan government did not abandon assessing educational progress in terms of student test scores. Constructivist principles of learning, it was found, led to a decline in test scores, and citizens began to realize that students lacked a knowledge of their cultural history and traditions. One observer noted that "young people interpret traditions, cultural heritage, and history any way they want." Realizing the failure of constructivist-based education, the government is now initiating new reforms that promote more directly the Western-consumer dependent lifestyle—which will further devastate an already degraded environment.

Japan's latest educational reforms, which start in the lower grades and are to be extended into secondary schools, are also based on constructivist-learning principles. But it appears that the transition from the fixed content curriculum that led Japanese students to score higher than students in other countries in the areas of mathematics, science, and language ability, is to be based more on a combination of Deweyean constructivism and traditional content that emphasizes learning Japanese moral norms, national and local history, and identity. The reforms are intended to foster what is called "zest for living" (*ikirurchikara*), and are to be less content driven and thus less of a source of pressure on students and parents. The explanation of what is to be achieved in constructivist-based educational reforms includes references to developing the student's sense of justice, awareness of responsibility, autonomy, ability to learn and think independently, and grow in his or her own individuality. Teachers are to achieve these contradictory goals by promoting independent learning through individualized instruction. To further enhance the self-directed learning of the students, elective courses are to be introduced in the fifth grade and later extended into the higher grades. As is the case in Taiwan, Japanese teachers and parents are also uncertain about the value of contructivist-based reforms. And teachers are unsure about how to implement these changes—particularly how to reconcile the autonomy of the learner with the need to pass on the shared cultural history, moral norms, and an appreciation for beauty and for the environment. There is also concern about whether the constructivist approach will lead to mastery of content that is essential to an increasingly technology-dependent society.

The introduction of constructivist learning into the classrooms of Albania, Turkey, and Pakistan, to cite just a few of the Islamic cultures that have welcomed in the Trojan horse of Western values and patterns of thinking, represents another example of the double bind facing tradition-oriented cultures that are attempting to modernize by adopting an approach to education that makes a virtue of cultural amnesia. Conflicts between tribal groups that are asserting their traditions and territorial rights in the face of the political boundaries imposed by the West are only one of the sources of turmoil in this region of the world. There are also conflicts that have their roots in the influence that Western secularism has had in recent years, as well as in the renewed efforts to recover the unadulterated teachings of the Qur'an—which is itself a source of controversy. But there is another challenge facing the regions inhabited by groups living in accordance with the teachings in the Qur'an, which is to obtain a clearer understanding of the relationship of the lifestyle that is required by Islamic teachings and ecological systems that are being placed under severe stress.

To the outside observer, it would seem that the introduction of educational reforms that promotes the student's self-discovery of knowledge would further exacerbate these tensions, particularly between a secular, individualistic-based lifestyle and a life based on the traditional customs rooted in the Qur'an. More specifically, it would also seem that the three dimensions of the Qur'anic way of knowing—metaphysical, naturalistic, and human—as they are translated into the daily practices that affect the commons, will become an increasingly marginal part of the students' lives. Concern about the eroding effect of the secular, consumer-oriented values promoted in American public schools has even led many American Muslims to homeschool their children when Muslim schools are not locally available. That they would base their homeschooling on the constructivist principles that foster the anomic individualism they see celebrated by the media and shopping malls is totally unimaginable. Their awareness of this fundamental contradiction raises the question of who is inviting the Trojan horse of constructivism into countries such as Turkey and Kazakhstan that are deeply rooted in Islamic traditions, and whose economic interests will be served if constructivist reforms are successful in creating the new form of individualism.

The government of Mexico provides yet another example of how constructivist-based educational reforms are being promoted as part of a national policy for promoting the Western model of economic development. But in the case of Mexico, many of the indigenous cultures are actively resisting what they perceive as yet another effort to undermine their traditions. For example, the Zapotecs and surrounding indigenous cultures in the Oaxaca region of Mexico are resisting the government-sponsored educational reforms. They are acutely aware that the modernizing educational reforms threaten to undermine their cosmic vision that requires harmonious relations with all the forms of life that share in the commons. And their response to this threat has been to withdraw their children from the government-sponsored schools. As Gustavo Esteva and Madhu Suri Prakash observe in *Escaping Education: Living and Learning Within Grassroots Cultures* (1998), parents are no longer accepting the idea that schools provide their children with a superior form of knowledge. They are, instead, recognizing that their knowledge, which has been refined over hundreds of years of experience in sustaining the commons as a life-giving ecosystem, is what really matters in the education of their children.

Participating in the rich symbolic life of the commons in these cultures provides children with a profoundly different experience and form of learning than is promised by constructivist-learning theorists. The following description by Esteva and Prakash needs to be considered in light of the dogma held by the constructivists that knowledge cannot be transferred—but must be discovered and constructed by the child. "In many villages," observe Esteva and Prakash, "the first birthday is celebrated when the child is three years old. That very day, accepted as a full member of the community, the child begins participating in most community activities births, deaths, feasts, funerals, and all the regular rituals of a rich community life." (p. 79). As one observer has pointed out, this is the beginning of the child's ancient future—that is, a future based on the traditions rooted in the ancient past and renewed in ways that help ensure the survival of the present and future generations.

In the next chapter I will examine how constructivist-based educational reforms contribute to the globalization of the Western model of development, so it is useful here to address why many Western readers (especially Western educators) are likely to dismiss as either romantic or evidence of reactionary thinking the argument that knowledge and values handed down from the past and renewed in the lives of the current generation is essential for maintaining a vital and sustainable commons. The difference between what Guillermo Bonfil Batalla calls "imaginary culture," which fits more the romantic and culturally uninformed thinking of Dewey, Freire, Piaget, Doll, Oliver, and their many followers, and these anciently rooted cultures is the ability of the latter to ensure the self-sufficiency of their communities. In answering the question of what the indigenous type of economy has to offer, Batalla points out that

> it offers basic security, a broader margin of subsistence in difficult years, even though one has only what is really indispensable. Various crops, together with wild plant gathering, hunting, fishing, and the raising of domestic animals, intermixed with some sort of handicraft production (pottery, textiles, basketry, and many other products), and the generalized capacity for other sorts of work such as construction and maintenance—all offer a broad spectrum of possibilities that can be altered or combined, according to the circumstances. No one of these possibilities alone, given the conditions of indigenous communities today, assures survival. Together, however, they offer an acceptable margin of security. For this multiple strategy to succeed, each activity must be on a small scale, producing what is neces-

sary and nothing else. This fact contains another general characteristic of the Indian economy: its low level of surplus and its low level of accumulation. 1996, p. 28

Batalla identifies another characteristic of indigenous communities found both in Mesoamerica and South America that is profoundly at odds with the constructivists' imaginary culture of emancipated, autonomous individuals. As he observes:

> The notion of salary is foreign to a large part of the work oriented toward self-sufficiency. Work is not paid, it is returned; one acquires the obligation of doing for others what they did for you, when the appropriate time arrives. Communal work is an implicit obligation of being part of a community. It is a universal obligation, without distinction. Here . . . when someone doesn't go, he must pay someone else to work for him. Taken together, these forms of cooperative work organize the efforts and abilities of the community according to the priorities that are decided by the community itself, or by its organized authorities. p. 31

An approach to education that fosters individual autonomy, as the above quotation should make apparent, is more suited to an economy that is dependent upon wages. Unfortunately, a livelihood based on wages is becoming even more vulnerable to economic forces that individuals and communities have little control over. And when wage-based work is not available there is increasing poverty and misery. In too many instances, instead of cooperation, there is competition for the few jobs that pay a wage.

How Constructivism Undermines the Commons in North America

The revitalization of the commons is integral to addressing the eco-justice issues that are increasingly being ignored as public schools and universities in North America align themselves more closely with corporate culture. In the United State, corporate logos, fast-food outlets in public school and university cafeterias, and the use of computers are now the most visible linkages. The forms of knowledge that universities have elevated to high-status—print and other abstract language systems, scientific inquiry, incessant drive to create new ideas and technologies, and assumptions such as anthropocentrism (which is also the basis of eco-management courses) mechanism, individualism, and the equating of a university education with a higher material standard of living—provide the necessary conceptual and moral framework that leads to a consumer-dependent lifestyle. As I have pointed out elsewhere (1997, 2001, 2003), what has been relegated to low-status by being omitted entirely from the curriculum or severely marginalized, are the forms of knowledge essential to the health of the cultural and environmental commons. That is, the non-monetized relationships, activities, and forms of intergenerational knowledge that enable students to become active participants in the cultural commons makes them less dependent upon consumerism. This, in turn, leads to a lifestyle that helps address eco-justice issues.

The five unresolved and too often unrecognized aspects of eco-justice that are directly related to our hyperconsumer lifestyle include the following: (1) the impact that the industrial toxins and wastes have on the health of marginalized social groups in society; (2) the disruption of local economies and traditions of self-sufficiency in third world countries that results from the exploitation of their resources in order to sustain our level of hyperconsumerism; (3) the destruction of the symbolic foundations of the commons that results in people being more dependent upon consumerism to meet daily needs—even as the opportunities for earning a living wage are decreasing; (4) the failure to limit the pursuit of materialism in ways that ensure that the quality of life of future generations will not be diminished; and (5) the right of other species to renew themselves and to not be reduced to an exploitable resource.

Addressing these eco-justice issues will require a radical reconsideration of how high-status knowledge perpetuates the process of economic and cultural colonization that accompanies the global spread of the industrial mode of production and consumption. In short, this will require that environmentally oriented teachers and professors will need to broaden their focus to include cultural patterns of self-sufficiency, the many ways in which science has helped us to understand the changes in the natural systems that are part of the commons, and the ways in which science undermines the symbolic systems of cultures that have a nurturing relationship with the commons. But addressing eco-justice issues is the responsibility of more than just the small number of teachers and professors who now take seriously the ecological crisis. The responsibility needs to be taken on by all teachers and professors—regardless of their discipline. Indeed, one of the most cogent recommendations that Rolf Jucker makes in *Our Common Illiteracy: Education as if the Earth and People Mattered* (2002), is that a transdisciplinary approach needs to be taken if students are to understand the cultural roots of the ecological crisis, and the cultural changes that will contribute to a more sustainable future (pp. 333–334).

If we examine the educational background of the people who promoted the Free Trade Organization, run the World Bank, and support the expansion of the North American Free Trade Agreement, we would find that they are mostly graduates of the country's most prestigious universities. And if we inquire into the educational background of the politicians who are now promoting the American imperial world order, as well as the heads of corporations that are supporting this project on the grounds that it will open up new markets for the steady stream of new technologies, we would find that few if any encountered professors who helped to clarify the connections between this imperialistic agenda and the destruction of the commons—both in North America and in other parts of the world. The connections between a university education and a hyperconsumer lifestyle, with its accompanying status symbols of a sports utility vehicle, an oversized house, and latest computer gadgets, would be relatively easy to research—that is, if sociologists and anthropologists were to become interested in how the failure to address eco-justice issues affects the commons. It would also be relatively easy to research the size of the ecological footprint of people whose lives are based on the forms of intergenerational knowledge and values that are currently excluded from the curricula of public schools and universities.

The Commons of Rural America

The commons in rural America represent a mix of rapid population decline and, in other regions, population growth. In rural areas where the Euro-American population is declining (eastern Montana, western and central regions of the Dakotas, Nebraska, Kansas, and the northern part of Texas) the natural systems are beginning to revert to the buffalo commons. In some areas the human population is as small as 2.5 persons per square mile. In other areas, the indigenous population is increasing—fulfilling an Indian prophecy about the interconnections between the return of the buffalo and the revitalization of their cultures. These changes in human demographics also affect the cultural aspects of the commons. Fewer people on the land has meant that towns have become smaller and less viable—with some being reduced to a grocery store, post office, gas stations, café, and a church. Attending a public school increasingly involves being bused to a larger population center. While the older farm population still has a social life centered around these meeting places and one that is increasingly reliant on television and the Internet, the youth of these shrunken towns are moving to the urban areas where opportunities seem greater.

In rural towns where the population has not diminished to the point where main street is largely an empty row of storefronts, the symbolic aspects of the commons can still be considered as resources that public schools can help students connect with. During my high schools years when I spent my summers working on a wheat ranch in eastern Oregon, I found that my urban biases kept me from recognizing the value of many small-town activities—such as the service associations, the long conversations that took place at the local café and the machinery repair shop, and the Saturday night dances in the nearby towns. The larger issue that transcended my urban biases is that these face-to-face conversations were the basis of the community's mutual support systems—as well as the source of important information about changes in the local environment, the impact of national economic forces on local farm prices, and the constant negotiation of relationships and obligations that needed to be reciprocated. Without getting into the question of whether the conversations that sustained the human web we call community were ennobling or the expression of seemingly unrelenting banality, it represented a non-monetized aspect of the commons that is vital to keep alive.

Writers such as Wallace Stegner, Kathleen Norris, Wendell Berry, Wes Jackson, and Stephanie Mills bring out the richness of rural landscapes and community-centered activities. The challenge for rural educators who understand the romanticized urban images, and thus their pull on youth, is to help students recognize the range of communal activities that are going on around them. Kathleen Norris' *Dakota: A Spiritual Geography* (1993), describes her return to a small North Dakota town after spending years in New York City. The spiritual geography she discovered was in the range of talented people she encountered—people who can serve as mentors to students with an interest in poetry, different visual arts, local theater, service activities, gardening, a wide range of crafts, and in rooting their lives in the natural ecosystem part of the commons. As all of these rural-oriented writers point out, the face-to-face relationships within rural towns involve a different and more deeply grounded form of moral reciprocity than in urban areas where so many relationships are centered on heavily monetized activities and relationships. Jayber Crow, the central character in Wendell Berry's novel of the same name, found social class and ideological differences in his small town existence, but he also found meaningful activities and relationships, as well as ways of providing a service that was useful to others. His life, while not fitting the American dream of upward mobility and material success, was meaningful to himself and others. In the end, his own insights into himself did not leave him with a sense of personal failure and lack of integrity—which stands in sharp contrast to the company man portrayed in Arthur Miller's classic *The Death of the Salesman* and in the more recent film *About Schmidt* who had all the seeming advantages of the urban commons.

That the rural commons tend to be more conserving in orientation presents another challenge for teachers. Helping students to understand that the rural form of conservatism is in part a response to an awareness, often reinforced by narratives of past periods of deprivation and failure, that the weather and world conditions can change in unanticipated ways—leaving the farmer without a crop to market or to market one that further undermines the farm as a viable economic enterprise. A conserving orientation also arises from years of experience of closely observing the changes in natural systems, which too often indicate a decline in the fertility of the soil, the lack of adequate moisture levels, and the further loss in the diversity of animal species. The conserving orientation can also be an expression of the farmer's sense of stewardship and desire to leave to the next generation the same prospects for a productive and meaningful life. But teachers also have another challenge, and that is to help students understand the economic and political forces that pressure farmers to use technologies that are environmentally destructive, and to go under economically. The corporate approach to agriculture, including the dangers of monocrops, genetically engineered seeds, and the use of chemicals, also need to be understood. The efforts of small farmers to use less environmentally invasive technologies and to establish interdependent relations with nearby urban centers, both of which contribute to the survival of small farms, should also be an essential part of the curriculum for the third of American students who are still educated in rural schools.

How to utilize the design processes in nature—in obtaining sources of energy, using building materials, siting buildings in ways that enhance the aesthetics of the landscape—is also an aspect of intergenerational knowledge that does not fit the constructivist emphasis on student-centered inquiry. Learning the principles of ecological design, which requires relying upon local knowledge of weather patterns and other physical characteristics of the land and animal population, and on the skills of local farmers, also needs to be supplemented with a knowledge of the new solar-dependent and energy-efficient technologies—which the rural teacher needs to incorporate into the curriculum. Along with helping the students acquire this knowledge, they also need to work with members of the community in providing students the on-the-ground experience of work that is not paid, but is returned by other members of the community when similar situations arise in the future. These understandings, as well as experiences as participants in adding to the commons, cannot be achieved by students constructing their own knowledge. Indeed, the curriculum that supposedly fosters individual autonomy and thus emancipation from intergenerational knowledge, would lead to a narrowing of the students' encounters with both the human and natural aspects of the commons. The influence of the corporate-controlled media and computer games on the thought patterns of students will have already predisposed many of them to leave for the more urban environment where entertainment could more easily be matched to their immediate and rapidly shifting interest. Rather than the one-sided emphasis of the constructivist teachers on changing the world, revitalizing the rural commons requires teachers who can bring together mentors that have a long and accurate memory of the land and the changes it has undergone as a result of the assaults of modern technologies, the ecologically informed experts who understand the conserving nature of new farming techniques, the people who understand both the economic global forces and the local economic needs of the community, and the students who need to be able to envision the range of communal, economic, and spiritual possibilities of being long-term members of the commons that sustained their parents—and will sustain their progeny.

◊ The Commons of Urban America

The commons in urban America, as Lewis Mumford pointed out decades ago, are dependent upon the commons of rural America, including those of third world countries. The air and the parklike settings are about all that is left of the natural systems in urban areas, and the air is heavily contaminated with health disrupting chemicals. The rest of nature's bounty—fresh water, sources of food such as vegetable and animals (increasingly grown in industrial-style settings), wood, minerals fashioned into building materials and other products, etc.—are largely imported from the rural commons. The huge volume of waste that is part of the urban lifestyle of consumption is, in turn, shipped to the less populated regions where its destructive impact on distant commons results in fewer protests.

Urban areas nevertheless can still be thought of as possessing the characteristics of the commons. In many instances, people share the language, narratives, ceremonies and foods of their ethnic group, as well as participate in many aspects of the symbolic commons of the dominant culture. Others share important metanarratives that are the basis of their religious or secular-based self-identity. In addition to the artists, writers, politicians, sports figures, and bureaucrats who plan and run the urban infrastructure, there are many skilled and selfless mentors who are willing to help others discover and develop their own talents—ranging from playing chess and artistic performances to writing and gardening. The range of activities and talents found in urban areas—even in the most seemingly blighted neighborhoods, is remarkable. More remarkable still is the spread of monetized activities, relationships, and products that are marginalizing what remains of the urban commons.

The activities and relationships within urban America have increasingly become commercialized. Theater, musical performances, health care, sports, communication (increasingly computer mediated), and even the preparation of food now require participating in the monetized economy. Unlike the commons in many indigenous cultures (and in communities such as Ithica, New York), where work is often part of a barter system, the urban commons has become so transformed by market relationships that many youths have little awareness or interest in the older generation that possesses the talents and ability to mentor them in ways that reduce dependency upon the monetized economy. In effect, the spread of consumerism—even into the area of human reproduction—has socialized urban youth into adopting a taken-for-granted attitude toward assuming that everything they do has a monetary cost associated with it. This narrowing of possibilities is furthered by the loss of work opportunities and by the increasing percentage of jobs that now pay a minimum wage. In effect, the impact of the industrial mode of production and consumption is creating three interrelated crises that have an increasingly destructive impact on the urban commons: the spread of poverty and disillusionment, the disjuncture between the diminished ability of people to buy basic necessities and the excessive production capacity that forces more production facilities to be relocated in areas of the world where wages are even lower, and the increasing rate of environmental degradation. These crises are further exacerbated by class and racial differences—with the poor and ethnically marginalized being forced to devote more of their time and energy to acquiring the basic necessities as well as to being forced to live in heavily contaminated environments.

The distinction between the monetized and non-monetized aspects of the commons is especially important for public school teachers and university professors to understand. However, before elaborating on this point, it needs to be pointed out that in everyday life the distinction is seldom clear cut. The aspects of the commons where youth are relying upon personal talents and mutually supportive relationships, such as playing a game or participating in more artistic activities, still requires buying athletic equipment and perhaps paying for public transportation. Similarly, the mentoring relationships, regardless of the activity, also involve certain essentials that are purchased rather than entirely produced by the participants. On the other extreme, being a consumer of a sporting or musical event, fast food, or health care represents the other end of the continuum where little personal talent and self-development is involved. Perhaps the difference can be seen more clearly in children's play that involves storytelling and outdoor games learned from older children, and children's play that involves video games and Barbie dolls. It can also be seen in the difference between a group of youths who develop their own musical talents and youths who are consumers of the musical talent of others.

The distinction between monetized and non-monetized aspects of the commons is still useful to consider when thinking about whether the curriculum leads to greater or less dependence on consumerism. The writings of constructivist learning theorists not only ignore the ecological crisis; they also ignore the human/symbolic aspects of the commons—and how the life-sustaining aspects of the commons are intergenerationally renewed. Learning who the mentors are in the diverse ethnic communities represented by students in the typical urban classroom is seen as inconsistent with the classroom where it is assumed that knowledge cannot be transferred—and where the goal is to teach students the scientific method of problem solving (Dewey) and to emancipate themselves from the knowledge of previous generations (Freire, Giroux, McLaren, Doll, and Oliver). Learning about the connections between an increasingly monetized everyday existence and the ecological crisis, and how the spread of the culture of consumerism contributes to poverty both here and in other countries is also to be left out of the curriculum on the assumption that the student's interests are the best guide to what is important to learn.

I will return to the pedagogical and curricular issues related to strengthening the non-monetized aspects of the commons in a later chapter. For now, it is important to consider the connections between the constructivist-based educational reforms being promoted in both Western and nonWestern countries and the ideology that underlies the globalization of the Western model of development. As many writers view the process of economic globalization as the latest expression of colonization, it is important to consider whether the constructivist approach to learning has less to do with fostering local democracy than with preparing the next generation to live in the Western model of a global monoculture. As pointed out earlier, one of the primary characteristics of constructivist approaches to education is that it fosters a condition of cultural amnesia that makes resistance less likely. But like so many of the assumptions underlying all the constructivist theories of learning, cultures renew themselves in many ways that are not understood by constructivist theorists. If and when youths understand how they are being mislead, they are likely to participate in local resistance movements that will further marginalize the need for cooperation in addressing ecological issues that cross many boarders.

 chapter 5

Constructivism
The Trojan Horse of Western Imperialism

Kirkpatrick Sale's observation about the industrial approach to production and consumption brings into focus what has been generally ignored by constructivist learning theorists who equate emancipation of the individual with gains in social progress, justice, and democracy. In *Rebels Against the Future: The Luddites and Their War on the Industrial Revolution* (1995), Sale describes how the intergenerational basis of community self-sufficiency had to be undermined in order to create the form of individualism that would be dependent upon the products of an industrial system in order to survive. As he put it:

> All that 'community' implies—self-sufficiency, mutual aid, morality in the marketplace, stubborn tradition, regulation by custom, organic knowledge instead of mechanistic science—had to be steadily and systematically disrupted and displaced. All the practices that kept the individual from being a consumer had to be done away with so that the cogs and wheels of an unfettered machine called 'the economy' could operate without interference, influenced merely by invisible hands and inevitable balances and all the rest of the benevolent free-market system. p. 38

As the constructivist learning theorists share the same taken-for-granted cultural assumptions with the classical liberal theorists who provided the ideological justification, including moral legitimacy, for the earlier and, now, current phase of the Industrial Revolution, they have not recognized how their ideal of the autonomous, critically reflective individual is essential to the industrial, capitalistic system they often criticize.

Sale's observation about the connections between replacing intergenerational knowledge of the community (the commons) with consumer products highlights another feature of the commons that needs to be more fully understood. As mentioned in the previous chapter, the earliest understanding of the commons was the pasture, streams, woodlands, animals, and so forth that the peasants shared and governed in terms of local norms and decision making. Also mentioned earlier is that the commons can be understood as having a symbolic (that is, cultural) dimension that encompasses everything from a common language, systems of decision making, technological knowledge, patterns of metacommunication, narratives, and so forth. In effect, the commons encompasses everything that has not been privatized. The commons may also include, depending upon the mythopoetic narratives and traditions of the culture, the shared ways of stratifying wealth and the exercise of political power. In order to avoid romanticizing the commons it is also important to recognize that it also includes the shared norms that govern market relationships—including the conceptual and moral norms that are the basis of colonizing the commons of other cultures. It needs to be recognized, however, that these shared norms and ways of thinking represent a perversion of the ancient idea of communal trust and intergenerational responsibility.

The idea of the commons as encompassing everything commonly shared now needs to be understood as far more inclusive than the earlier understanding of the shared resources and responsibilities connected with the bioregion. Our understanding of the commons can be rescued from the confusion that may accompany the broad inclusive definition given above by introducing another concept—which is that of "enclosure." The first legal and ideologically based assault on the commons of England occurred in 1235 with the Statute of Merton. The legal justification for enclosure, which was the code word for privatization, was that it would lead to improvements and thus to extracting greater rent (*The Ecologist*, 1993, p. 23). Since that time the concept of enclosure has not only been associated with privatization of the land and its resources, but with the commodification of labor, ideas, and everything else that is now being organized and controlled by an industrial process. When enclosure was extended to how work was shared within the community, it lead to reversing the idea that work is returned, to work is paid—at the lowest possible level. Enclosure, in effect, can be understood as the exercise of power that privatizes, stratifies, marginalizes, excludes, and shifts the decision making process from the local community to outside centers of power. The multiple processes encompassed by this seldom used word are now disguised by the god words of "development," "modernization," and "progress."

But the logic of industrialization, and the new forms of enclosure that it dictates, are not the only way of understanding the commons—or what remains of it. There are still aspects of the commons that have not been privatized and integrated into the industrial approach to markets. It is these aspects of the commons that need to be understood in order to consider how the educational process can be used to address eco-justice issues as well as contribute to what Vandana Shiva refers to as "earth democracy." Where the cultural groups are knowledgeable about the self-renewing natural systems they depend upon, the commons is sustained through the practice of local democracy. The modern expression of enclosure, on the other hand, usually involves a shift in the locus of decision making to corporate offices and, now, to institutions such as the World Bank, the World Trade Organization, and international treaties that regulate trade. The difference between local decision making about what is shared in common and the political process that accompanies various forms of enclosure can be seen in how the former leads to keeping the focus on the common good—for the members of the human and natural community. An example of decision making in a community that understands that the unit of long-term sustainability is the local ecosystem that the human community is dependent upon can be seen in how the people of the Wabigoon Lake Ojibway Nation of Ontario, Canada, still harvest the wild rice as part of the commons. Even with the introduction of machines, the harvesting is still regulated through community decision making. Where machine harvesting may be carried on, and where the traditional canoe method of harvesting is more appropriate are decided by the community. Limits on the size of the harvest for both approaches are set, and violators may be denied harvest rights for the rest of the season (*The Ecologist*, 1993, pp. 13–14).

Jeannette Armstrong, a member of the traditional council of the Penticton Indian Band of British Columbia, brings out another way in which the revitalization of the commons is understood and practiced. In addition to understanding that "self" is relational and that actions thus affect the other participants in the local ecosystems, Armstrong points out that the Okanagan language has been derived from what the land has taught previous generations. She writes that

> We . . . refer to the land and our bodies with the same root syllable. This means that the flesh that is our body is the pieces of the land come to us through the things the land is. The soil, the water, the air, and all the other life forms contributed to our flesh. We are our land/place. Not to know and celebrate this is to be without language and without land. It is to be dis-placed. 1996, pp. 465–466

Many Anglo- and Euro-Americans, as well as ethnic groups from other parts of the world, exercise local decision making in ways that help to revitalize the beliefs and values as well as the other material aspects of the commons most directly related to their sense of identity and group memory. Local decision making may even extend to a wide variety of communal interests: the use of water, prohibitions against the use of certain chemicals, small cooperative efforts, nurturing of the arts, preservation of recreational areas, and so forth. Local decision making may even extend to the creation of a local currency that enables the members of the community to exchange services and local products in a way that represents the community's decision about what represents fair value. But not all of these examples of local decision making that address common interests are based on the same awareness that is found among many indigenous cultures, which is that humans must practice moral reciprocity and earth democracy with the non-human world. Rather, most of the nonindigenous cultural groups in North America who practice environmental stewardship still view themselves as separate from the land—but open to learning its lessons. As Gary Snyder notes in *The Practice of the Wild* (1990), most Western people still associate nature with wildness, and thus as fundamentally different and separate from humans—even as our body and consciousness interact with and replicate the basic life-sustaining processes found in other "wild" organisms. The sense of being separate from and superior to nature, leads to our continual effort to subjugate it by technologies that will have long -term consequences that few people consider and even fewer care about. And what cannot be subjugated or privatized is set aside as wilderness areas—that is, as national parks (pp. 8–18).

The enclosure of the commons now takes new forms of expression, such as the privatizing of the airwaves and, now, even what is most distinctive about humans—the digitizing of thought and communication. Private ownership is easily recognized as a form of exclusion, as is the pervasiveness of the corporate control of the media that filters the news and orchestrates consumer demand. But modern technologies are less well understood as expressions of enclosure and exclusion. The assembly line, as the Luddites recognized, marginalized the workers' craft knowledge, control over their own time and pace of work, and the sense of personal satisfaction that accompanies producing something useful for others. Experts who must first create a sense of fear and limitation in order to create the illusion that their services are superior to the non-monetized intergenerational knowledge that sustained patterns of mutual assistance, and thus was part of the commons, now occupy nearly every niche in mainstream society. This self-interest masked as altruism further marginalizes the local systems of self-sufficiency and governance. Megastores such as Wal-Mart and industrialized-food outlets such as McDonald's also are expressions of how market forces represent new forms of enclosure.

What remains of the commons is being further undermined by computers that are being represented as connecting people in ways that makes communication near instantaneous, more efficient, and worldwide in its reach. This technology has been acclaimed as contributing to democracy and enabling people to form cyberspace communities. It has made many more things possible, and has enabled complex systems to be more easily predicted and controlled. The many benefits can-

CONSTRUCTIVISM: THE TROJAN HORSE OF WESTERN IMPERIALISM | 83

not be denied. But the computer is also part of the enclosure process that turns more of the commons into monetized relationships. Now thought and communication have been brought more directly into the market economy. Seldom recognized is how computers contribute to a process of enclosure that marginalizes and excludes. Like the other expressions of enclosure cited above, computers amplify certain cultural patterns while excluding others—or transforming them to fit the cultural patterns of decontextualized data and information that it amplifies. This process of amplification and reduction (marginalization and exclusion) can be seen in the language-processing characteristics of computers. Like all language systems, technologies that mediate language simultaneously illuminate and hide. Even the word "computer" illuminates and hides. What is illuminated and thus privileged is the idea that the mind functions like a computer—processing data and information, and constructs ideas and plans of action based on supposedly objective data. To make this point somewhat differently, computers amplify both the advantages and disadvantages of print—and thus contribute to a literacy-based form of consciousness. What is marginalized by computers and other print-based texts are the many forms of expression that face-to-face communication can take. The latter cannot be digitized without being fundamentally transformed into an abstract text that requires the same form of individualism Sale described as essential to the success of the industrial approach to production and consumption. As computers become more dominate in mediating thought and communication, they contribute to the dynamics of enclosure where people cease to remember the importance of face-to-face communication to mutuality and moral reciprocity—which are essential to membership in the commons.

What is generally overlooked as computers become the icon of modernity and the cutting edge of technological development are the many exclusions that result from their mediating characteristics. Like the telephone that amplifies voice and, now, pictures over vast distances, while reducing and eliminating the non-verbal patterns of communication, the computer has its own built-in cultural agenda of enclosure which can be understood in terms of what it amplifies and reduces. The computer's role in the process of enclosure can most easily be understood by considering what cannot be digitized without being taken out of context. The list of enclosures include the different cultural and physical contexts of face-to-face communication, the moral norms that are specific to different cultures and that govern the patterns of interpersonal and environmental reciprocity, the embodied forms of knowledge rooted in experiences of place and interactions with other forms of life, the narratives and ceremonies that are (if told truthfully) the basis of self-identity and continuity with the past and future, the mentoring relationships that form character and bring forth previously unrecognized interests and talents, and the mythopoetic narratives that are specific to different cultural groups and that are the basis of how members of that group understand their relationships with each other and the natural environment. Also marginalized is an awareness of the metaphorically layered language that carries forward the culture's past ways of thinking and value system—leaving people, as Gregory Bateson points out, with outdated conceptual maps that still influence what aspects of the environment will be recognized and how they will be interpreted.

In addition to continuing the process of enclosure carried on by today's public schools and universities, where the encoding process privileges the abstract and thus context-free knowledge over the spoken and thus context dependent word, computers also amplify the cultural patterns that undermine the culturally diverse approaches to the commons. The symbolic systems that are the basis of how different cultural groups have successfully adapted to the commons on a sustainable basis are always local and dependent upon reciprocity that is both intergenerational and interspecies. To be more specific, the conduit mode of communicating data and information that appears as free of cultural and physical context, of human authorship, and of culturally specific ways of knowing, reinforces the myth that words and other abstract symbol systems refer to an independent and nonculturally mediated reality. The printed word and other abstract systems of representation are, as cognitive scientists would have us believe, the basis of thought—that is, the stuff that enables individuals to construct their own ideas and plans of action. To make this point another way, computers amplify the Western myth that represents thought as an attribute of the autonomous individual.

Computers also reinforce the Western assumption that equates change with progress, and that the monetization and thus expansion of markets into more areas of the commons is the way of measuring progress. The form of individual subjectivity reinforced by computers further strengthens the dominant Western assumption that the individual is separate both from nature and from traditions. In terms of the latter, the computer creates the illusion that the individual exists outside the flow of temporality. That is, as a decision maker about where to connect in cyberspace, and with whom, the individual also decides whether traditions and the future are to be considered at all. In short, computers undermine the importance of individual and communal memory, as well as the awareness that in this era of rapid change traditions need to be constantly assessed in terms of their importance to community and in terms of their destructive impact. Given the vast amounts of data, information, and expert knowledge that can be quickly accessed, the importance of mentors and elders recedes even further from awareness.

The amplification and reduction characteristics of computers should not be interpreted as meaning that they should be abandoned (which would now be an impossibility even if it were decided that the losses far outweigh the benefits). Rather, what needs to be considered is the relationship between the cultural context and the reason for using the computer. The computer has proved its usefulness in gathering information on changes taking place in natural systems, but its use undermines the commons when it is used as the chief means of interpersonal communication. It is useful for scheduling the arrival and departure of airplanes, but its impact on young children, as the accumulated evidences shows, is more damaging than beneficial. In every situation, there is the question: What are the benefits and losses—and who benefits? Do local communities benefit when the use of computers makes it possible for corporations to outsource the production process and thus jobs to other regions of the world where wages are lower? Does the use of computers transform more aspects of the commons into market relationships—thus increasing the need for money that is increasingly difficult to obtain in an environment where computers are being used to reduce the need for workers? Context, cultural differences, and the intergenerational knowledge and norms of reciprocity that are essential to the health of the commons are important considerations in deciding what constitutes the appropriate uses of computers.

To summarize, the role that computers play in the process of enclosure include the following: they monetize the most basic attributes of humans—thought and communication; they marginalize the importance of face-to-face intergenerational communication that is essential to maintaining what Gary Snyder refers to as "the contract a people make with their local natural system (1990, p. 31); they privilege abstract and reductionist thinking over embodied and relationally derived knowledge; they exclude the mythopoetic narratives that connect people's lives with the larger symbolic/spiritual world that for many cultures includes other forms of life; they privilege the subjective judgment of the individual in a way that makes the empowering and hard-won traditions that communities are based upon contingent on the individual's subjective perspective and mood; and they further undercut the already tenuous awareness in Western culture that life needs to be lived in ways to ensure that future generations will find a sustainable environment.

It may appear that the many ways in which computers undermine the commons is a tangential issue that is unrelated to constructivist-based educational reforms. This would be an incorrect way of thinking about computers. In the *Report to the President on the Use of Technology to Strengthen K-12 Education*, written by the President's Committee of Advisors on Science and Technology, the justification for recommending that $13 billion should be spent annually on expanding the classroom use of computers was that students would be better able to construct their own ideas more effectively if they could use computers to access data. The report also recommended that constructivist principles of learning should be more widely implemented in classrooms even though, as the report acknowledged, these principles remain unproven (1997, pp. 135–136). Ignored by the President's Advisory Committee is that the data, information, and simulation software programs designed for student use encode the cultural assumptions of the people who collect the data and write the simulation programs. That is, their report ignored how the student's use of computers is part of a process of socialization where the language that appears on the computer screen often becomes the basis of the student's taken-for-granted way of thinking—as described in the earlier chapter on the cultural basis of intelligence.

In addition, there is another connection between constructivism and the classroom use of computers: namely, the ways in which both the various interpretations of constructivist learning and the classroom use of computers contribute to the globalization of a Western mind-set, with its emphasis on an industrial approach to production and consumption. The processes of marginalization and, in some instances, total exclusion of the different ways in which face-to-face interactions are the basis of the community's mutual support systems can be seen in what cannot be digitized without being transformed into an abstract printed text that fits one of criteria of what constitutes high-status knowledge in the West. It can also be seen in how the reliance on computers in the increasingly wired world is further undermining all but a few of the most widely used languages. There is yet another way of understanding how constructivism and computers are contributing to a global monoculture that will further stress the viability of natural systems—and further transform the world's climate systems in ways that could have a catastrophic impact on life as we know it. And that is to consider how the cultural assumptions shared by the various constructivist-learning theorists are also shared by the developers and promoters of computers—and how these same assumptions underlie the current efforts of transnational corporations and international institutions such as the World Trade Organization to extend capitalism into every city and village in the world.

⊠ Constructivism as an Ideological Trojan Horse

There are several ways of understanding the nature of an ideology. Marx understood an ideology as a set of beliefs that created a false state of consciousness where the higher values and beliefs extolled by the ideology hid the existing patterns of exploitation. Clifford Geertz, the cultural anthropologist, views an ideology as the set of beliefs, assumptions, and values that serve as a cultural template that governs everyday life (1973, p. 216). His interpretation of how an ideology functions does not exclude the possibility that the prevailing system of assumptions and values may privilege certain groups within the culture over others. On the other hand, Geertz does not equate all ideologies with false consciousness. An ideology, such as the cosmovision of the Quechua and many other indigenous cultures, may be the basis of moral reciprocity with all other participants of the commons. In other cultures it may be the basis of class divisions and exploitation, as well as the basis for treating the environment as an exploitable resource.

An ideology may also be the driving force and conceptual blueprint for reforming society, and for bringing economic development and Enlightenment ways of thinking to supposedly less advanced cultures. In this sense, the ideology is assumed to be based on a more progressive and emancipatory way of thinking. Its other characteristics include a missionary zeal to reform the entire world, an unquestioned sense of altruism, and a deep conviction that its mission is based on a "Truth" that only rational, progressive thinkers are able to grasp. As a blue print for reforming our own culture as well as others, it may be an example of Marx's understanding of an ideology that hides the dynamics of exploitation and centralization of economic and political power. The condition of false consciousness may even include viewing the environmental crisis as unrelated to the marginalization of ethnic groups.

How an ideology can also become the basis for revitalizing the commons, maintaining cultural diversity, and addressing eco-justice issues also needs to be briefly explained. Ecologically centered cultures are based on belief systems that combine Geertz's way of understanding an ideology as a cultural template with Snyder's understanding that life within the commons is a contract between the generations and the natural community. At the center of this contract is the shared understanding of the importance of moral reciprocity, intergenerational knowledge of how to live in sustainable ways, and earth democracy—that is, recognizing the rights of plants, animals, and other natural systems to regenerate and govern themselves. As a set of guidelines for reforming Western culture that is now addicted to a lifestyle of hyperconsumerism and the exercise of power over natural processes, it should lead both to raising awareness of the reactionary assumptions that lead to degrading natural systems and to colonizing other cultures—often in the name of spreading democracy, individual freedom, and progress.

How constructivist-based educational reforms are socializing students to accept the same classical liberal and thus reactionary assumptions that are shared by transnational corporations and institutions such as the World Bank and the World Trade Organization (WTO) can be seen by recalling the cultural assumptions that underlie all the constructivist-learning theorists and comparing them with the assumptions that underlie the transnational corporations and international institutions they have created for the purpose of overturning the policies of national governments that impede free trade. To recall the assumptions that are shared by the major constructivist theorists, they all represent *change* as the dominant feature of everyday life. Thus, the educational goal is to promote critical reflection, experimental inquiry, autonomous thinking associated with logico-mathematical reasoning—all of which assume that the traditions that are the basis of different cultures are irrelevant. If the student should become aware of traditions, his or her task is to reconstruct them. Constructivist theorists, with the exception of Dewey, emphasize that the one true method of thinking they advocate leads to *individual autonomy*—which involves existing in a state of continual becoming. All of the constructivist theorists promote human-centeredness; that is, their approach to educational reform is *anthropocentric* in that the individual's role (or social group for Dewey) in promoting progress does not need to take account of the way natural systems are being degraded. Lastly, all of the "fathers" of constructivist learning theories represent their one-true approach to knowledge as being more evolutionary advanced than other ways of knowing. They are all *universalists* (which is another term for colonizing thinkers) in the sense of advocating that their respective one-true approach to knowledge should be adopted as the standard for the rest of the world's cultures.

These shared assumptions are the basis of the ideology that is now being used to transform the world's supposedly backward and diverse cultures into a world monoculture based on the Western pattern of thinking. Before comparing the constructivist blueprint for changing the world with the blueprint (ideology) of transnational corporations and the world-governing institutions such as the World Bank and the World Trade Organization, it would be useful to summarize what is excluded in the constructivist-based educational reforms being introduced in countries around the world. As students construct their own knowledge and direct the course of their own education, the following will be ignored and thus excluded from their education: (1) an awareness of differences in cultural ways of knowing, including an understanding of how different cultures disrupt or nurture the commons; (2) an awareness that cultures encode and intergenerationally renew knowledge in different ways—and that youth in these diverse cultures learn from different cultural sources and from different activities; (3) an awareness that cultural diversity is essential to maintaining biodiversity; (4) an awareness of the different ways in which cultures maintain a balance between market-related activities and the non-monetized relationships and activities of everyday life; (5) an awareness of how the traditions of different cultures contribute to patterns of moral reciprocity and mutual aid, as well as traditions that are sources of privilege and marginalization; and (6) an awareness that the two most important questions that students need to ask are: What needs to be conserved that contributes to the well-being of the commons? What needs to be changed—and how can it be done in a way that involves the fullest participation of the community?

How the forces promoting the globalization of a Western model of an industrial- and consumer-dependent lifestyle share the same ideology that underlies constructivist-based educational reforms can be seen in the agenda of the World

Trade Organization. The WTO is the creation of misnamed "conservatives" groups—corporations, chambers of commerce, and governments led by politicians indoctrinated by their university education to accept the ideas of late eighteenth- and early nineteenth-century classical liberal thinkers as having the same universal validity as the law of gravitation. That John Locke, Adam Smith, John Stuart Mill, and, later, Herbert Spencer, were addressing the problems of their day, and were not knowledgeable of other cultural ways of knowing, has been totally disregarded by those who have appropriated their ideas for the purpose of justifying the authority and agenda of the WTO. Governments around the world have ceded to the WTO the right to use its rule-making and rule-enforcing power to override decision making at the local, state, and federal level in all countries that are judged to be impeding free access to markets and the right of corporations to maximize their profits. As Debi Barker and Jerry Mander note in their monograph, *Invisible Government* (1999), the WTO has been given the legal authority to enforce the "massive transfer of economic and political power away from national governments and into the hands of global corporations and the trade bureaucracies they create" (p. 1).

Established in 1995 by agreement of 134 countries, housed in Geneva, Switzerland, and with a staff of over 500, the WTO rules on complaints brought by corporations and governments (generally acting on behalf of corporations) about restrictions on free trade. The perceived restrictions may take the form of local, state, or national environmental and labor-protection laws, tariffs, and other regulations designed to protect health standards and local economies. Decisions about what constitutes a barrier to free trade and the right of corporations to conduct their operations without local restrictions are made by the WTO's Dispute Settlement Body, which is made up of a three-member panel of corporate and trade lawyers. If the decision goes against the government or corporation that brings the complaint about the restriction of free trade or about a government that is attempting to protect local and national interests threatened by the practices of a corporation, the decision can be appealed. The appeal is then reviewed by a seven-member Appellate Body, and its decision can only be overturned by a unanimous vote of the current 144 member nations. Both the Dispute Settlement Body and the Appellate Body conduct their hearings in secret.

The WTO also possesses the authority to enforce its decisions, even when it threatens a local economy, the health of a nation, and traditions essential to cultural identity. The enforcement may take the form of economic penalties and trade sanctions against the losing party. As Barker and Mander point out, "The loser has three choices: change its laws to conform to WTO requirements, pay permanent compensation to the winning country (or corporation), or face harsh, permanent trade sanctions from the winning country" (p. 8). The threat of these penalties has led may countries to adapt their laws in ways that will prevent them from being challenged by corporations.

Two examples of WTO decisions will provide an understanding of how the WTO places the interests of corporations above those of local and national governments. In 1998 and again in 1999, the United States, on behalf of food exporting corporations, challenged the Japanese regulations that set the standards for pesticide residues on produce that would be allowed into the country. The WTO ruled that Japan's health requirements were higher than those deemed necessary by the WTO, and that Japan was therefore in violation of international trade rules. In effect, the WTO ruled that Japanese citizens must accept a level of pesticide-contaminated food that was higher than their own government deemed to be safe. The case the United States brought against the Canadian effort to include more Canadian content in Canadian magazines sold at Canadian newsstands by giving tax incentives to Canadian advertisers was decided in favor of the United States. Canada was required to rescind the favorable tax and postage advantage that was intended to stem the tide of American influence on Canadian culture.

The WTO now has the authority to regulate the flow of money across national boarders and to decide on cases involving intellectual property rights. Indeed, its ability to eliminate any national or local barrier that limits the ability of corporations to pursue their economic interest, including the use of technologies that threaten the health of local inhabitants, means that the various systems of government, as imperfect as they are, are now superceded by the WTO. Countries have been required to compensate corporations on the profits they would have made if the local climate of opinion or law had not forced them to set up their operations elsewhere. The WTO, as well as other institutions designed to enforce a level economic playing field for corporations, are now including water resources, schooling, and other public-supported services on their list of barriers to the expansion of capitalism as the dominant mode of economic activity.

The agenda for the future, which the WTO shares with the World Bank and neo-liberal politicians, is to privatize what remains of the commons. That is, their goal is the complete enclosure of the commons. This goal, which harks back to the "survival of the fittest" Social Darwinian thinking of the late nineteenth and early twentieth century, is now being given new legitimacy by scientists such as Richard Dawkins, E. O. Wilson, and Gregory Stock. And it is a major threat to cultural diversity, to the billions of people who already live impoverished lives, and to sustainable ecosystems.

Protecting what remains of the commons in different parts of the world is being made even more difficult as corporations now have the legal authority and economic resources for overturning democratic decision making at the local level. The media, with its relentless messages that connect consumerism with individual success and happiness, now cross the boundaries of different language groups, states, and bioregions that are the commons of different cultural groups. Corporations promoting an industrial approach to food, medicine, agriculture, entertainment, communication, and schooling, are also free to impose themselves on the commons wherever they see potential markets—or exploitable resources. The spread of Wal-Mart across the country is an example of how corporations can weaken the economic vitality of small communities—which, in turn, reduces the face-to-face interactions on Main Street that are important to a sense of belonging to a community of shared interests. The spread of superstores and shopping malls needs to be understood as the industrialization of the commons—which requires, in turn, the autonomous individual who must work in order to consume what the media prescribes as evidence of personal success.

The ways in which the commons in other countries are being threatened can be seen in the way the World Bank, as a precondition for a $25 million loan to the government of Bolivia, required that the municipal water system be privatized. When a subsidiary of the Bechtel Corporation took over the water system in the city of Cochabamba, the cost of water was raised from 35 to 300 percent. Even community and private wells were made subject to the new water fees. As the cost of water began to exceed the cost of food for many citizens the government was unable to ease the crisis as the World Bank had stipulated that no part of the $25 million loan could be used to reduce the cost of water. Following massive protests and strikes, martial law was declared in Cochabamba. The directors of Bechtel and its local subsidiary then fled the country, taking with them their documents and computer files. This victory of local democracy, however, did not mean Bechtel will be the loser. As Maude Barlow reports, Bechtel is suing the government of Bolivia for close to $40 million, claiming that the government impeded the corporation's rights to conduct business and to earn a profit. The claim against the government was taken to the World Bank's International Court for the Settlement of Investment Disputes (2001, pp. 24–25). In effect, the violation of the commons instigated by the World Bank will be settled by one of its institutional branches—thus sending the message that local democarcy has no real safeguards against the global reach of corporations and the transnational institutions they have created to enforce their will.

Local resistance, however, has not been entirely overwhelmed. Local farmers in different regions of the world have resisted the introduction by Monsanto of the terminator seed that was genetically designed to produce a non-replantable seed, thus requiring farmers to purchase new seeds every year. Resistance may, in fact, be growing. In Nigeria, the petroleum industry is being challenged by local residents for fouling their water and land. Peasants in India are resisting being displaced by the building of dams, the privatization of water, and the industrialization of agriculture. And in Central and South America resistance is focused on the actions of oil companies, the industrialization of agriculture, and the disruption of local economies by free-trade agreements. Many other indigenous cultures such as the Quechua and Aymara resist being brought into the industrial model of development by continuing their traditional practices of earth democracy that have sustained them for over eight thousand years. The rapid spread of a more ominous and violent form of resistance to the Western project of globalization can be seen in the growing number of followers of Sayyid Qutb, the most important Islamic revolutionary thinker of this century. Of his many books, *Milestone*, has become the manifesto of the terrorist wing of the Islamic fundamentalist movement that is now being guided by Osama bin Laden. I mention the Islamic terrorist movement here because constructivist-based educational reforms are being promoted in Islamic countries where their Western notions of individualism and democracy are not likely to be seen as contributing to the recovery of the glory days of seventh-century Pan-Arabism.

The suggestion that the cultural assumptions that constructivist thinking is based upon are also the cultural assumptions, with several exceptions, that underlie the ideological blueprint that is being used by transnational corporations and their international legal enforcers is likely to be rejected by the followers of Dewey and Freire. The most vehement objections will come from the Freirean professors who also view themselves as in the Marxist tradition of being critics of capitalism. The constructivist followers of Piaget and von Glasersfeld (the theorist who claimed that "knowledge cannot be transmitted") are not likely to have an opinion either way, as they are so focused on promoting the students' own constructions of knowledge that the ecological, economic, and political changes taking place around the world are irrelevant to them.

As I do not wish to explain again the key assumptions and mode of knowing advocated by the different constructivist theorists that support the ideology of the transnational corporations, I will instead summarize how the constructivist-based educational reforms create a greater dependence on the market while at the same time undermining the intergenerational knowledge that represents, within different cultures, the community-level sources of resistance. The emphasis that each constructivist-learning theorist places on the student using one mode of knowing (critical reflection, experimental inquiry, reliance on subjective judgment and experience, and the openness to "becoming") as the basis of constructing their own knowledge about the world contributes to a fundamental disconnect between youth and the intergenerational knowledge that co-evolved with changes in the local bioregion. Again, it must be emphasized that not all the cultural forms of intergenerational knowledge contribute to standards of social justice that most of the world would agree with, nor to sustaining the viability of the local commons. However, it also needs to be recognized that the symbolic basis of moral reciprocity, the local systems of mutual support, skills, and knowledge that reduce reliance on a market economy and the practice of earth democracy are not likely to be derived from the student's construction of knowledge. The earlier constructivist approach to classroom learning in the nineteen twenties, which has been called the child-centered phase of the progressive-education movement, did not result in students becoming aware of gender bias, the clear cutting of forests, and the environmentally destructive agricultural practices that led to vast amounts of topsoil being blown away during the early years of the Depression. Nor did the students become aware of how their thought processes and self-identity were influenced by the culture's root metaphors that were encoded in the language they used in taken-for-granted ways.

Rather, the sources of resistance to reliance on the industrial approach to production and consumption, which are expressed in the ability to be more self-reliant as a community, are handed down and renewed in ways that take account of changes in the commons—as well as changes resulting from outside influences. This process takes place through mentoring, personal observations, face-to-face interactions, embodied experiences, narratives, questioning, insights into alternative ways of doing things, and participating in ceremonies that transform the ordinary daily experience by connecting the participants to a larger sense of purpose. Professors who write about the need for teachers to become "transformative intellectuals" and to teach (indoctrinate) students with the idea they should question everything have misunderstood that genuine resistance is not in listening to the teachers who urge them to question everything, but in being able to repay a work obligation through the use of a skill learned from others, to engage in a conversation, to tell a story, to play an instrument, to mentor youth, to pass on knowledge that heals, to repair some aspect of material culture, to read what the environment is communicating about its cycles of renewal and the ways in which it is being stressed, to prepare and share a meal, and so forth. Genuine expression of resistance means becoming less dependent upon the industrial-prescribed lifestyle. Resistance also involves acts of affirmation of the relationships that sustain the commons. These community-centered patterns of self-sufficiency are also the basis of local democracy—as Robert Putnam tells us in *Making Democracy Work: Civic Traditions in Modern Italy* (1993).

If the members of the community have become so individually centered that they do not know what traditions are the basis of mutual self-sufficiency and moral reciprocity, they will have few references or resources for resisting the relentless indoctrination of the corporate-controlled media. They will then be more dependent upon consumerism to meet both material and psychological needs, which, in turn, will pull them into the cycle of needing to keep up with the latest technology and other consumer fads. Being caught in this cycle leads to more debt, and the need to work more in order to avoid falling further into debt—which an increasing number of people cannot avoid. The condition of dependency upon consumerism involves becoming caught in another double bind where the ability to keep consuming and to paying off the accompanying debt is being undermined by the logic of the corporate approach to the development of new technologies—which is to create new markets while at the same time reducing the need for human workers who are more expensive to maintain than robots and computers. The absolute absurdity and misery associated with this drive to create new markets by introducing technologies into the production process that require fewer workers can be seen in the slums that surround most cities in the nonWestern world—and in the inner cities in many Western countries. The shops and store windows may be filled with every kind of manufactured product, but fewer people have the money to buy them. Thus a system that contributes to poverty by destroying the basis of personally fulfilling community-centered work, through its ever present visual reminders of what "successful" people are able to acquire, also communicates that people too poor to buy what is available are deficient as human beings.

Sale's observation about the need to destroy the traditions of self-sufficiency and moral reciprocity within communities in order for the Industrial Revolution to succeed represents another point of convergence between the constructivist-learning theorists and the ideology that is now used to justify transnational corporation's being held accountable only to the Darwinian law of survival of the fittest. Freire and his many followers who promote emancipation from all of the cultural practices of the previous generation unwittingly perpetuate the double bind that characterizes the global spread of the industrial approach to production and consumption. While they view themselves as critics of capitalism and all forms of exploitation, their efforts to make emancipation an ongoing process, with no consideration of what needs to be conserved, promote the form of individualism that will be more dependent upon consumerism. At the same time, their recommendation that there is only one approach to knowledge, that is, critical inquiry, undermines the many ways in which communities intergenerationally renew themselves. The followers of Freire exclude the learning of skills that contribute to individual and community self-sufficiency, and they exclude the importance of the many forms of learning that are based on face-to-face relationships. Thus, the double bind connected with their approach to constructivist learning is that in undermining the community networks of support and learning, their approach to education contributes to the kind of individual who will need a job that requires no special skills, which means working in an industrial setting that is driven by the need to create technologies that require fewer workers.

In making this criticism I want to emphasize again that I am not saying that all community networks, practices, and moral norms should be beyond criticism. Rather, the criticism is that Freire and his followers do not recognize that there are many mutually supportive traditions and networks in communities—even in those communities that may treat unfairly certain groups of people. Unjust and marginalizing practices should be the focus of criticism and reform efforts. But the daily practices that represent mutual-support systems and are alternatives to being dependent upon consumerism should be integrated into the educational process. Not only does the Freirean approach lack a sense of balance as well as a set of criteria for identifying what needs to be reformed and what needs to be conserved and intergenerationally renewed, but the followers of Freire have been extremely effective in preventing the publication of books critical of their view of education as a process of perpetual emancipation—as I have learned in my attempt to find a publisher for a collection of essays by leading third world activists who tried to implement the Freirean approach to teaching literacy in a variety of indigenous contexts and found that his pedagogy was based on Western assumptions that had a colonizing effect.

The World Trade Organization, and the transnational corporations whose interests it represents, have no interest in whether the traditions of different cultural groups are oppressive or contribute to morally coherent and relatively self-sufficient communities. All traditions that impede the widespread acceptance and, over time, dependence on new technologies are to be overturned. The computer and the cell phone are good examples of the addicted consumer who is unable to avoid purchasing the next technological upgrade—the next fix that feeds the addiction and that adds further to personal indebtedness. This is the kind of tradition that is reinforced by the industrial ideology, and it is radically different from the traditions that have developed over many generations of living within a particular bioregion, and that have their roots in a mythopoetic narrative that provides a way of understanding the purpose of life as something more than being a consumer in an industrialized world of constant change. And just as the ideology that promotes the global spread of an industrial based lifestyle is totally silent about the importance of the commons, and the degradation of the ecological systems that it depends upon, the Freirean approach, as well as the other approaches to constuctivist-based learning, share the same silence.

It needs to be pointed out that their unacknowledged complicity in furthering the ideology that underlies the global design of corporations and the WTO is not unique to these educational theorists. Richard Rorty, whose importance as a philosopher led to his being invited to present at University College, London, and Trinity College, Cambridge, the lectures that became the basis of *Contigency, Irony, and Solidarity* (1989), presents a view of tradition that is essentially the same as that of Freire and his current followers. Just as Freire represents traditions as being passed on only through a "banking approach" to learning, Rorty refers to traditions as the final vocabulary—that is, something fixed and beyond questioning. Rorty also views the ideal form of individualism in the same way that Freire and his followers do—that is, as constantly renaming the world through a process of critical reflection. The ideal for Rorty is what he refers to as the "ironist" individual who lives in a society where the only social consensus should be "to let everybody have a chance at *self-creation* to the best of his or her abilities" (p. 84, italics added). His description of existing without traditions (which he grossly oversimplifies as a final vocabulary) is also worth noting. As he explains the possibilities of life in the ideal liberal society:

> The ironist spends her time worrying about the possibility that she has been initiated into the wrong tribe, taught to play the wrong language game. She worries that the process of socialization which turned her into a human being by giving her a language may have given her the wrong language, and so turned her into the wrong kind of human being. But she cannot give a criterion of wrongness. p. 75.

I must admit that when I learn that 75 percent of Americans support a war that is questioned by a majority of the world's population and whose promotion was based on a series of gross misrepresentations, I also wonder whether I was born into the wrong tribe. But I can give a number of criteria for resisting the idea of preemptive wars. The problem with Rorty's ideal, which he shares with Dewey and Freire, is that he does not understand the complex and culturally varied nature of traditions—and how we re-enact them even as we declare that we can exist without any traditions. Rorty not only carries on the tradition within Western philosophy of ignoring other cultural ways of knowing, but he also carries forward the philosopher's traditions (now undergoing a slight change) of anthropocentic thinking. Even Rorty's reliance on the conventions of print-based communication are traditions that he subscribes to even as he represents them as constraints on the self-creation of the ironist individual. Perhaps the most important point to be made about Rorty's reactionary form of liberalism is his total dismissal of intergenerationally connected communities as sites of resistance to the destruction of the commons. He can also be faulted for his romantic view of the self-creating individual—a romantic view that Dewey expresses when he claims that people should experience everyday life as a constant process of participatory decision making that leads to the never-ending process of reconstructing experience. The Freirean theorists also share this romantic view of individuals constantly engaging in critical inquiry and renaming the world—even though, as in the case of Rorty's ironist individual, the criterion for renaming the world will itself be subject to constant criticism and renaming. The social engineers working to shape public opinion to accept the latest technological innovation use their critical capacities in the service of a clear set of criteria—and that is to find new ways to expand the market for what the techno-industrial system is geared up to produce. The task of these social engineers is made easier by the way in which these constructivist-learning theorists, along with classroom teachers who freely mix the different genres of constructivist ideas with their own romantic notions about being a facilitator of the student's self-creation, have undermined the students' awareness of how they are embedded in the networks of mutual support that constitute what remains of the commons. The experience of being alone and responsible for self-creation in a world of constant change is exhausting—and many students will find it easier and more relaxed to go to the local industrial-food outlet than to create their own recipes, to turn on the television or computer than to engage in a conversation or engage in self-creation out-of-doors, and to consume other people's music rather than to create their own.

Constructivist-based educational reforms are being promoted around the world on the grounds that they will foster the kind of individual that is able to engage in democratic decision making. Indeed, the connection between the student's self-construction of knowledge and a democratic society is taken for granted by all the constructivist-learning theorists—and by all the classroom teachers who try to avoid "transferring" knowledge out of a concern that they are thwarting the students' ability to think for themselves. Understanding that the Western individually centered approach to democracy has both strengths and weaknesses, and that it is not the only approach to participatory decision making has not been a strength of any of the constructivist learning theorists. Their arguments for their respective one-true approach to knowledge is evidence that they did not recognize the right of other cultures to maintain their traditions of participatory decision making— which varies in terms of how a cultural group decides the qualifications of the people who can best represent the interests of the entire community—and sometime the interests of the commons. Indigenous cultures in North and South America practice participatory decision making that does not fit the Anglo- Euro-American model of an individually centered democracy. And if we looked at group decision making in Bali about how water is to be allocated to the rice paddies, we would find another expression of local and direct democracy. There are other cultures where the political process is dominated by a hierarchical system of power relationships, and the interests of the entire community are secondary to the privileged class. The issue is thus not one of arguing for the abandonment of democracy, but rather whether we have the right to impose our view of democracy on other cultures. There is something decidedly undemocratic about our tradition of imposing our assumptions and political system on other cultures—particularly when we make so little effort to understand their belief systems and processes of decision making. And there is another important issue about exporting our approach to democracy that is being increasingly compromised by the ability of corporations to influence the politicians who are, in theory, supposed to represent the interests of the people that elected them.

This issue has to do with whether there is a basic flaw in the idea that democratic decision making is determined by individuals whose votes reflect their own self-interest. By way of contrast, we have seen the democratic process work in behalf of the common interests. Environmental legislation, as well as legislation protecting the safety of workers, are just a few examples where the democratic process has reversed practices that threatened the well-being of the entire society. However, it is possible to cite a greater number of examples where the democratic process has been subverted by powerful interests groups. The question that needs to be addressed here is the constructivists' claim that their approach to fostering autonomous, self-creating individuals reinforces the possibility of democratic decision making—and that their approach to educational reform should be exported to the world's other cultures. Individuals who have been reinforced in thinking that their own experiences and subjectively determined knowledge should be the basis of their political decision making simply perpetuate the most shallow way of understanding the democratic process—where political decisions are based on the greatest number of votes. Whether the greatest number of voters have made informed decisions or decisions that reflect a deep understanding of the moral issues simply does not matter. And then there is the problem that when people do not have the background knowledge, or exhibit the trait Rorty admires, which is to not have any criteria that serve as a moral guide, they are more susceptible to the reductionist, image-shaping political ads on television.

All of the constructivist-learning theorists make a virtue of not exposing students to knowledge in any systematic way. Rather, it is the student's shifting interests, stage of cognitive development, and, for the followers of Freire, the political agenda of the teacher who is to act as a "transformative intellectual" that determine what the student encounters in the way of curriculum. By excluding the traditions of the community, which vary from culture to culture, that contribute to social and ecological justice, that are sources of individual and community empowerment, and that represent alternatives to being dependent upon consumerism, students receive an education that conditions them to accept the continual changes that are introduced by new technologies. Exercising critical reflection, which has a legitimate role in the process of democratic decision making, needs to take account of what should be conserved as well as what needs to be changed. The constructivist theorists ignore the possibility that a critical understanding might lead to resisting change—such as the response we should be witnessing as the use of computers is now being used by the government to increase its powers of surveillance of what citizens are doing, and the response that should have been made to the adoption of international treaties that create institutions such as the WTO. The education of the voter who understands the threat of these and future changes needs to be more grounded in a knowledge of the network of community support systems, in the traditions of social and ecological justice that have been intergenerationally renewed and are now being threatened, and in ways that foster a sense of responsibility for improving the prospects of future generations. The educators who embraced the use of computers in the classroom on the basis that it will enable students to access information and thus to think more effectively are examples of the problem I am identifying here. That is, most professors of education and classroom teachers did not understand how the computers would affect relationships, undermine skills in interpersonal communication, and the many community-centered forms of knowledge that would be displaced as computers were integrated more directly into the industrial process of production. They were thinking of change as the expression of progress and individual empowerment—which is in line with the ideology used to justify a constructivist approach to classrooms.

There is yet another convergence between the cultural assumptions reinforced in constructivist-based classrooms and the ideology that is being used to eliminate local and national self-governance in order for transnational corporations to pursue their economic self-interest without any restrictions. This threat to local democracy, which takes different forms of expressions in different cultures, is not the main point I want to make here—but it is related in an important way. And it needs to be emphasized again that local decision making, which is essential to protecting what remains of the commons from unrelenting efforts of corporations to maximize profits, is being overturned by agreements made in secret by WTO bureaucrats in Switzerland. Unlike North America, where religion is still considered a matter of individual choice, many of the cultures where constructivist-learning principles are being introduced as part of a modernization agenda are based on deeply held religious beliefs. For many of these cultures, religion is their cosmology—that is, their world view that governs every aspect of daily life. The North American approach to religion maintains a clear separation of church and state; which is a separation that is becoming increasingly blurred as some religious groups in the United States are exerting more pressure on all levels of government to pursue policies that are in line with their interpretation of the Old and New Testament.

Constructivist-learning theorists are secular thinkers who carry forward the tradition of representing religion as an individual matter. While they are not overtly hostile to religion, their emphasis on students constructing their own values as well as their emphasis on treating change as overcoming the authority of traditions should be viewed as undermining it. Dewey is the only constructivist theorist who addressed the question of religion, but like the theory of experimentally based knowledge that he thought should be adopted by all the world's cultures, he did not understand the nature and role of the world's varied religions. In effect, Dewey thought that the values promoted in the Bible could be justified on pragmatic grounds. His followers have simply bypassed any discussion of the connections between a culture's religious cosmology by giving more attention to Dewey's dictum that values should be determined on a democratic basis. And as the individual, according to Western thinkers, is the basic social unit, the starting point in democratic decision making about what values are to guide relationships and provide an overall sense of purpose should begin with the individual.

Thus, the Trojan horse of Western thinking can again be seen in this aspect of constructivist-based reforms being introduced into Islamic and Buddhist cultures, and into indigenous cultures such as the Quechua. To cite just two examples of religious cosmologies that are being undermined both by constructivist approaches to learning and by the efforts of the World Bank to create what they call a "level playing field" that will enable transnational corporations to eliminate the competition of local producers, both the Islamic and the Quechua cosmology (which are profoundly different) represent moral/spiritual frameworks that are opposed to making economic relationships and the pursuit of profits the highest moral endeavor. Insofar as constructivist-oriented classroom teachers are able to indoctrinate students in these cultures to view themselves as autonomous thinkers and authors of their own values, which I think is far more difficult than the promoters of the constructivist ideology assume, they are undermining the deepest levels of symbolic resistance that these cultures have to being overwhelmed by the materialistic and individually centered culture of the West. As Helena Norberg-Hodge documents in *Ancient Futures: Learning from Ladakh* (1991), the combination of Western media exposure, the marketing of Western technologies that requires becoming part of a money economy, and the exposure to a Western style of education, succeeded in transforming a Buddhist culture that was essentially self-sufficient and rich in wisdom about human/nature relationships into a society where there is now poverty, unemployment, and intergenerational alienation. This process of undermining the religious/moral basis of cultures is part of the process of adopting the Western approach to modernization. While modernization may lead in some instances to reforming religiously based cultural traditions that are sources of extreme cruelty, it also undermines more broadly the moral basis for resisting the spread of consumerism and the poverty that accompanies the industrial system of production.

Constructivism's contribution is to cloak its secularizing influence behind what is represented as the higher values of modernization necessary for becoming a member of the global community: democracy, individualism, progress, science, and freedom. Unfortunately, these values, as least in the Western context, have not always been understood or practiced in ways that have safeguarded the commons for future generations. Nor have they been practiced in ways that have led to a more eco-justice-based society. Given the size of the ecological footprint of the average American citizen, the American model of education, which has been based on various interpretations of constructivist learning principles, should not be the model that other cultures should be encouraged to follow. We view ourselves as the best-educated and most individually responsible society, yet there has been nearly total silence on the part of the public about how the WTO is undermining what remains of local democracy—and only a minority of the citizens have expressed concerned about the integrity of the commons.

Given how constuctivist-based educational reforms are contributing to the process of globalization and thus to further degrading the environment by the individually centered lifestyle it promotes, it is necessary to sketch the outline of alternative approaches to education that address the problems ignored by the constructivist theorists. The main oversights include how to help students reconnect with the values and practices of their culture that contribute to maintaining the life-sustaining capacity of the commons, how to revitalize the traditions that represent sources of resistance to the seemingly overwhelming influence of Western technology and materialism, and how to understand the double binds inherent in Western knowledge systems that promise technological empowerment while undermining local traditions of self-sufficiency. It is hoped that the following chapter will provide a starting point for a dialogue centered on these issues and for initiating fundamental reforms in teacher education that take these issues into account.

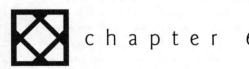

chapter 6

Toward a Culturally Informed Eco-Justice Pedagogy

The principal concerns that should guide the reform of teacher education, and thus the education that occurs in classrooms, have been articulated by several writers who have kept in focus the connections between the cultural patterns that are degrading the environment and the rise of political conflict, as well as the spread of poverty around the world. In the last chapter of *Our Common Illiteracy* (2002), Rolf Jucker discusses 27 reforms that will contribute to transforming classrooms from sites of indoctrination into the culture of consumerism and environmental exploitation into sites for learning how to live in ways that contribute to a sustainable future. His recommendations range from incorporating into the curriculum examples of ecologically informed approaches to development in other parts of the world and making education for a sustainable future the core feature of every course, to providing staff development in how to teach an ecologically informed curriculum.

David Orr has also recommended several basic reforms that will contribute to ecological literacy. In addition to suggestions for how to contribute to the student's knowledge of place, as well as his recommendations for making school buildings more energy efficient, Orr's insights into the pathology of American culture that teachers need to avoid perpetuating express a Wendell Berry-type of wisdom. As he writes in *The Nature of Design* (2002):

> Ecological design at the level of culture resembles the structure and behavior of resilient systems in other contexts in which feedback between action and subsequent correction is rapid, people are held accountable for their actions, functional redundancy is high, and control is decentralized. At the local scale, people's actions are known and so accountability tends to be high. Production is distributed throughout the community, which means that no one individual's misfortune disrupts the whole. Employment, food, fuel, and recreation are mostly derived locally, which means people are buffered somewhat from economic forces beyond their control. Similarly, the decentralization of control to the community scale means that the pathologies of large-scale administration are mostly absent. Moreover, being situated in a place for generations provides long term memory of place and hence of its ecological possibilities and limits. pp. 9–10

His summary of the basic relationships that are too often ignored can be interpreted as a set of guidelines for teachers in how to renew the commons and local democracy. While some readers may interpret Orr as suggesting that we should return to an earlier, less technologically driven lifestyle, a more accurate interpretation would be to view these recommendations as practical steps for resisting the technological and economic forces that are subverting what remains of local democracy and are increasing the economic vulnerability of local communities.

In effect, Orr is reiterating the danger of becoming increasingly dependent upon an industrial system that is not accountable in terms of the needs of local communities, the limits of local ecosystems, and the prospects of future generations. Orr's comparison between the virtues of settled, ecologically informed cultures and the consumer-oriented values that many Americans base their lives upon, even while they claim to be the chief custodians of conservatism, could also serve as a checklist for teachers who take seriously whether they are perpetuating the problems of the commons—or contributing to its renewal. What the industrial system has turned into the higher values of our age, according to Orr, is "showiness, ego trips, great wealth, huge homes, hurry, excessive consumption." What has been lost, with the exception of small self-conscious communities in rural and urban areas, are "cooperation, neighborliness, competence, thrift, responsibility, and self reliance." These different value orientations are represented in the mix of students in nearly every classroom and at nearly every level of the educational process. The teacher's curricular decisions are thus critical to which set of values and cultural patterns become the focus of inquiry and affirmation and which become part of a silent process of indoctrination.

Educational reformers who address eco-justice issues in an increasingly conflict-ridden world need to take seriously the recommendations of Jucker and Orr. They also need to take account of the inequities and marginalization that has been part of American history—as well as the history of other Western-colonizing countries.

However, the solution to the problems of racism, class inequities, and gender marginalization must be understood in ways that do not strengthen the economic forces of globalization and the consumer-dependent lifestyle it requires. Promoting the industrial approach to production and consumption, along with automation and the outsourcing of production facilities to the low-wage regions of the world, is not the solution to overcoming poverty and unemployment. Indeed, this approach is showing itself to be a major contributor to the crisis, as we are seeing in American and Mexican communities where production facilities have been moved overseas.

A culturally informed eco-justice approach to teaching and the curriculum cannot, by itself, reverse the trend of globalization and cultural domination. It can, however, play an important part in challenging the deeply held patterns of thinking; and it can contribute to an understanding of the alternatives to the current emphasis on individual self-centeredness, the incessant drive to get ahead at the expense of others, and the spread of cultural amnesia that accompanies the increasing dominance of the new trinity of the military, the federal government/industrial complex, and the religious fundamentalists. The recent role of the school and university in raising awareness of the many expressions of racism and gender bias and in providing the knowledge and skills necessary for careers that previously could not have been imagined by these marginalized groups, is proof that the classroom can have an ameliorative effect. But first, teachers and professors must realize that they are complicit either as perpetuators or as problem solvers.

An eco-justice approach to educational reform also confronts classroom teachers and professors with an existential choice—of choosing to be aware that the degraded condition of the environment and the global spread of poverty, now widely reported in the media, may be related to the ideas and values they promote in their classrooms. They also make choices about whether they will inform themselves about the issues and problem solving approaches of their more ecologically informed colleagues. To put the problem in its simplest terms, the choices facing Western classroom teachers and professors about how to address the deepening ecological crisis is critical to averting the future that approximates Thomas Hobbes' prediction of human existence becoming reduced to "continual fear and danger of violent death"—a life that is "solitary, poor, nasty, brutish, and short." As Jared Diamond documents in *Collapse: How Societies Choose to Fail or Succeed* (2005), Hobbes' prediction has proven all to accurate for members of cultures that destroyed the habitat they depended upon.

As we are just beginning to understand the pedagogical and curricular issues related to addressing the unresolved eco-justice issues and to renewing the commons, the following guidelines should be viewed as provisional. The guidelines should also be adapted to fit the different cultural contexts. Cultures that already have a small ecological footprint and are now being pressured to adopt Western technologies and an industrialized economy will find that the guidelines must be interpreted differently. Their challenge will be to protect the traditions of the commons that do not adversely affect other cultures or natural systems. For these cultures, the guidelines need to be interpreted in a way that helps the students to understand the culturally transforming effects of Western ideas, values, and technologies. Teachers in Western classrooms will find the guidelines useful in addressing a different set of problems. The main one being how to renew the commons in ways that reduce the cycle of excessive consumption that leads to higher levels of accumulation and dispersal of toxic wastes that, in turn, adversely affect marginalized groups. This cycle is dependent upon the further exploitation of the resources of third world countries in order to raise even higher the level of excessive consumption.

What is common in both Western and nonWestern cultures is the teacher's role as a cultural mediator. Even in cultures where there is an effort to recover the moral norms dictated by a religious cosmology that goes back thousands of years, the globalizing of Western images that equate consumerism with happiness and individual success, as well as the presence of technologies based on Western science, confront teachers with the challenge of mediating between different cultural orientations. The teacher who is located in an indigenous community, where the classroom is itself an expression of Western influence, also faces the challenge of mediating between the traditions of the local community and the various expressions of Western culture. Nor can teachers in the West escape mediating between different cultural choices, with one set of choices being to perpetuate the high-status knowledge that supports the industrial system of production and consumption, while the other set involves strengthening the self-reliance of the community in ways that reduce wasteful consumption. Indeed, the question of which cultural practices and values contribute to an ecologically sustainable future by reducing the violence that accompanies economic and technological globalization confronts every teacher and professor—regardless of grade level or subject area. Choosing one set of cultural possibilities over others is a process of mediation that is inescapable—even when the teacher or professor is in denial.

There are many public school and university classrooms where the curriculum challenges the current global trends and introduces students to cultural practices that are being informed by ecological ways of thinking. However, the majority of public schools and university classrooms continue to perpetuate the assumptions and values that are the source of the double bind where progress is equated with an environmentally destructive dependence upon consumerism. And there are other double binds now endangering the entire world: the strident ethnocentrism of high-level American politicians in an era of religiously based radicalism and friend/enemy politics; the relentless drive to create new technologies that further eliminate the need for workers while at the same time creating new demands for the increased number of consumer products. There is also the threat hanging over the entire world of extremists pursuing their goals by using the technologies created by scientists who view themselves as modern Promethian heroes—but in the employee of the state. These double binds represent alternative futures for students. They also represent how classroom teachers and professors mediate by reinforcing the cultural assumptions that perpetuate these double binds, rather than helping students learn about the many ways in which ecological thinking can be applied in everyday life.

The more immediate double bind facing classroom teachers and the education professors who now socialize them is the belief that what students learn is incidental to the more paramount concern, which is that students "use what they know to interpret new information and construct new knowledge." As Sam Hausfather summed it up, "content . . . does not come first. Students' experiences, ideas, and prior knowledge come first" (2002, p. 77). For the followers of Freire, who continue to ignore both the ecological crisis and that the promotion of one approach to knowledge will further undermine cultural diversity, the teacher's task is to contribute to the humanization of the student by encouraging them to question oppressive traditions—which from the student's perspective could be anything and everything. There is another unrecognized double bind in these different interpretations of a constructivist approach to learning: namely, that the issues, ways of thinking, and uses of technology related to the ecological crisis are so much a part of the student's and teacher's taken-for-granted patterns of thinking that it would never occur to them that they should be made explicit and examined in terms of their impact on the commons. The taken-for-granted status of the cultural patterns of thinking that are ecologically problematic and the romantic vision of a global monoculture that is constantly undergoing changes initiated by teachers functioning as "transformative intellectuals (to recall Giroux's formula) have also resulted in the continued disregard of the most important issues facing both Western and nonWestern cultures. Just as the destruction of the self-reliant capacities of communities and the enclosure of the commons were essential to the growth and spread of the Industrial Revolution, the current lack of an in-depth knowledge of the shaping influence of technology, of the limits of scientific explanations, of the community-centered alternatives to consumerism, of the traditions essential to achieving eco-justice, and of how the principles of ecological design differ from the mechanistic models of thinking that are being applied in so many areas of everyday life, are essential to today's corporations, politicians, and media experts who are imposing the nineteenth-century classical liberal economic agenda on the rest of the world.

While the followers of Dewey and Freire assume, as Peter Roberts naively put it, that "all aspects of reality are constantly changing" (2000, p. 49), and the interpreters of Piaget merge his ideas with the earlier child-centered practices of the progressive education movement, there is a radically different way of understanding the role of the teacher. With globalization spreading to even the remotest corners of the world, the teacher mediates between the traditions of the community and the forces of change, between the short-lived experiences and the questions of students and the knowledge (even wisdom in some instances) accumulated over generations of living in a bioregion, between Western technologies and technologies that are adapted to local conditions, between Western science and intergenerational and place-based science, between the industrial model of production and consumption and living within the possibility of the commons, and between cultural pressures that reinforce subjectively based values and traditions that locate the source of values in a culture's formative mythopoetic narratives or religious traditions. The tensions, sources of threat, and supporting patterns will vary within and between different cultures. Regardless of whether they are in Western or non-Western classrooms, teachers cannot avoid mediating between the cultural cross currents and pressures of different cultures. But they may not always be aware of what cultural orientation they are privileging or how they are undermining the conceptual and moral basis of the commons.

Before discussing the teachers' responsibilities in their mediating role I want to digress briefly by answering the critics of my writings. The criticisms have taken many forms and are made by educational theorists whose ethnocentrism has led them to ignore learning about nonWestern cultures. One of the criticisms is that I take a culturally relativist position. As the existence of nearly 6,000 spoken languages, and thus cultural ways of knowing, still exist today (with many on the verge of extinction) I am not sure exactly what would be the opposite of taking the position that knowledge and value systems are relative to different cultures—and that within these cultures their knowledge and values (or the assumptions they are based upon) are taken for granted and are thus not experienced as relative. The opposite position, which my critics want me to adopt, would be the view that there is a universal moral and knowledge system—which is the position taken by many Western philosophers who have a long tradition of ignoring the existence of other cultural ways of knowing. As I have pointed out earlier, the idea that there is only one way of knowing (Freire's critical and culturally transforming inquiry, Dewey's method of experimental intelligence, Rorty's ironist individuals who doubt they were born into the right language community, today's classical liberals who think everybody should live by the rules of capitalism, and so forth) reflects a colonizing mentality. Indeed, there is no shortage of messianic thinkers in the West.

A second general criticism of my views is that I fail to provide the criteria for determining which traditions should be renewed and which should be reconstituted or abandoned entirely. One critic even went so far as to suggest that I failed to provide the criteria that other cultures should use to determine which of their traditions should be kept or abandoned. The suggestion that I should decide for others would be similar to the position that Dewey and Freire took and that their current followers now take. That is, the assumption that we have the right to impose on other cultures our supposedly more advanced way of knowing. This expression of hubris is not a unique trait of Western educational theorists. E. O. Wilson, Richard Dawkins, and a host of other prominent scientists and technologists have made similar proposals—such as Wilson's claim that all the world's cultures should adopt the theory of evolution as their guiding metanarrative. These nonWestern cultures are expected to embrace the theory of evolution that is being re-worked by scientists in a way that explains why they are being selected for extinction (Bowers, 2003). While I continually explain that the West can learn from ecologically centered cultures but that we should develop our own ecologically sustainable traditions rather than assuming that we can borrow from other cultures, I am represented as suggesting that we should go back to an earlier time by adopting their traditions—which is a criticism that not only misrepresents my writings but also reflects an evolutionary way of thinking that represents the West as having evolved to a more advanced stage of culture. A fourth criticism that is made by the followers of Freire is that I romanticize traditional cultures, which is really an ideologically based criticism that ignores that I continually refer to specific indigenous cultures. The criticism that my romanticizing of traditional cultures takes the form of representing them in only a positive way is the Freirean way of dismissing the need to take actual cultural differences into account—particularly differences that relate to their ability to live within the limits of their commons. In a recent article, Peter McLaren and Donna Houston even suggested that my discussion of indigenous cultures amounts to an attempt to revive the romantic image of the Noble Savage (McLaren and Houston, 2005).

In explaining the teacher's mediating role, as well as suggesting the content of the curriculum that should not be left to the student's powers of self-discovery, I will be restating, as I have done elsewhere (2001, 2003b), the connections between local democracy, intergenerational knowledge of the limits and possibilities of the commons, and the need to resist the forces of economic and technological colonization. These connections can also be viewed as one of the reasons I do not support the prescriptions of my ethnocentric critics who are also in deep denial about the nature and extent of the ecological crisis. And since these critics support their state of denial by claiming that indigenous cultures have also altered and degraded their environment, I need to reiterate what I have written before. Yes, many indigenous cultures have altered their environments in ways that Western scientists are just now beginning to understand as examples of good ecological management techniques. And yes, some indigenous cultures have undermined the life sustaining characteristics of their bioregions or faced environmental changes they could not successfully adapt to. That these cultures have ceased to exist is yet another lesson we can learn from.

It is also necessary to restate that critical reflection, experimental inquiry, dialogue between teacher and students, drawing upon student experience and interest, hands-on problem solving, and embodied ways of knowing are all essential to learning. They represent basic common sense. My criticism is directed at those who assume that there is only one legitimate approach to knowledge. The mediating role of the teacher needs to be based on a common sense understanding of the interplay of the social context of learning, the student's interests and level of background knowledge, what represents the most appropriate approach to learning (embodied, explanations, inquiry, ethnographic-based, etc.), and the cultural patterns that the teacher needs to make explicit. The teacher also needs to take account of the cultural differences in patterns of metacommunication, as I explain in *Educating for Eco-Justice and Community* (2001). But the mediating role of the teacher requires that other concerns that are marginalized by all the constructivist learning theorists be taken into account. These concerns have to do with the content of the curriculum that is ignored because of the constructivist emphasis on being a facilitator. Freire described this facilitator role as being "engaged in a continuous transformation through which they (the students) become authentic subjects of construction and reconstruction of what is being taught, side by side with the teacher, who is equally subject to the same process" (1998, p. 33). I quote Freire again because I want to represent him and his following as accurately as possible in order to highlight their suspicion that teachers who bring background knowledge to the learning situation are engaging in the banking approach to education.

This emphasis on teachers engaging students in a process of learning (experimental inquiry, critical reflection, becoming, etc.) creates another double bind that needs to be acknowledged. The double bind is that the teachers who are trained in the various constructivist schools of thinking often lack the background knowledge that is necessary to their role as cultural mediators. The result is that both the teachers and students do not know what they do not know. When this illusory state of consciousness is combined with the liberal ideology of constructivist theorists— where change is equated with progress, the student's subjective knowledge is viewed as the expression of autonomy, the relativity of values and knowledge as expressions of the student's "authenticity," and so forth—it is assumed that the students will be more self-realized, authentic, and free. Unfortunately, this is the state of consciousness that a consumer culture promises to raise to an even higher level of self-gratification. Contrary to Freire's messianic approach to engaging everyone in the ongoing process of renaming the world, which Roberts describes as "a continuous effort to reinterpret reality" (2000, p. 146), the traditions that might serve as sources of community resistance to the Western project of monetizing every aspect of daily life—in all cultures—are either ignored or subject in Maoist fashion to "continuous transformation" (Freire, 1998, p. 33).

◊ The Teacher as Cultural Mediator: Guiding Principles

The role of the mediator in labor negotiations is to bring together the two opposing sides and to facilitate a civil discourse that will lead to an agreement over differences. Just as the mediator in a labor dispute needs to understand the background of the two groups, thinking of the teacher as a mediator retains the same idea of bringing together different perspectives and interests. It also includes the idea that the teacher, when acting as a mediator, understands the historical, conceptual, and lifestyle patterns of the different cultures that come into contact in a classroom—in the lives of the students, through the use of curriculum materials, and the culture that the teacher brings to the classroom. Unlike the labor mediator whose purpose is to resolve the differences between two contending parties, the teacher's role is to help students understand the assumptions, mythopoetic narratives, economic systems, uses of technology, and, more generally, the impact that the different cultures encountered in the classroom have on the commons. Unlike the labor mediator, the purpose is not to achieve a new level of consensus, but to engage students in an examination of cultural differences—as well as what is shared between cultures. This understanding is essential to being able to recognize the likely consequences of adopting the traditions of another culture. In addition to the emphasis on clarifying cultural differences, the teacher differs from the labor mediator in another way. That is, the teacher's role is not to be neutral. How can the classroom teacher or professor mediate the differences between the agenda of the WTO and local decision making that may lead, for example, to a law prohibiting the privatization of the municipal water system? As we all know, the ability of teachers to mediate between political and economic entities is almost zero. However, if the teacher's mediation is to help students to understand the differences that separate two or more institutions, governmental policies, or powerful economic trends (such as outsourcing of production or allowing a corporation to pollute the local water supply)—including the historical roots, interests in terms of who gains and who loses, and the long term consequences for maintaining a life supporting environment—then it may lead to a more informed level of political action. That is, mediation in this sense may help to revitalize local democracy. And local democracy is the only real basis for resisting the forces that are undermining the commons—and the only basis for affirming which local traditions are vital to what remains of the community's traditions of self-sufficiency and identity.

The rate and scale of environmental degradation in the name of social progress should lead to abandoning the idea that the teacher can adopt the neutral stance of a facilitator of the students' decisions about what they want to learn. The idea that teachers should present both sides of an issue and then let the students make up their own minds has been based on a number of misconceptions that supposedly fair-minded people have embraced. However, what is often left out of the conflict where two sides of an issue are presented has often been more important than what was included in the class discussion. The Great Books approach to a liberal education, along with the other well-intentioned approaches to letting the students decide, did not lead to the discovery of racism or gender bias—until political action outside the classroom forced teachers and professors to take a position. After becoming aware of what had been previously ignored, many teachers and professors began promoting an understanding of equality between different groups. Those who clung to the old silences and prejudices were also taking a position, while those who became aware of what previously was taken for granted presented students with examples of racist and sexist practices, provided an understanding of the origins and historical developments that led to today's awareness of yesterday's denials, and engaged students in a discussion of social and personal consequences. Unfortunately, there are other silences that still outweigh in significance the two-sides-of-an-argument approach that still prevails in many classrooms. For most classroom teachers and professors, environmental problems are viewed as the responsibility of science teachers. And the rapid changes in cultures and local communities caused by the relentless efforts to monetize all aspects of life and the increasing automation of production and outsourcing are not even recognized—except where there is a glimmer of awareness that the increasing reliance on computers may eventually be automated to the point where fewer teachers will be needed.

Mediators in labor disputes often helped to improve worker safety, job protection, and salaries. But their neutrality was achieved by ignoring the need to take on the responsibilities of an educator. For example, they do not see their task as that of informing the public about the long-term implications of the trend that began nearly a century ago with decision making being taken out of the hands of the workers and made a function of management. Nor do they see their responsibility as educating the public about the economic and political agenda of neoliberal groups or, for that matter, helping the public understand what it is that politicians and corporate groups who call themselves conservatives really want to conserve. While the teacher's role as cultural mediator cannot bring two or more opposing groups together, they can provide the background knowledge about the history, interests, and future consequences that will likely result—depending upon which group prevails or what has become lost through lack of attention.

As there are few classrooms in the world today where Western influences are not present (if not in the classroom then in the community), I want to suggest that the following guidelines may be useful in clarifying the difference between the teacher's role as a cultural mediator, the teacher's facilitating role in a constructivist-oriented classroom, and the teacher's role as an agent of the continuous social transformation that Deweyian and Freirean theorists advocate. In suggesting the different curricular decisions that are part of the teacher's mediating responsibilities, I want to emphasize that I have no suggestions for how the local cultures should go about the educational task of renewing their own intergenerational traditions. Therefore, the following suggestions for thinking about the teacher's role as a cultural mediator are meant to apply to classrooms where different Western cultural traditions are present—in terms of ideas, values, behaviors, clothes, uses of technology, and so forth.

◆ First Guiding Principle

The teacher needs to be aware that the student's ability to exercise communicative competence in the political sense where language and background knowledge are essential to the determination of what issues are raised and future consequences are made explicit is dependent upon how the teacher mediates the process of primary socialization.

As explained in chapter three, when students are learning something for the first time (about some aspect of science, the use of a technology such as computers, or a historical fact or event, etc.) the words and concepts that are the basis of the new understanding exert a powerful influence on what and how the student will think about that aspect of the culture. For example, if the explanation of technology is limited to a vocabulary that conveys the concept of technology as a neutral tool (and as the expression of progress) most people will continue to think of it in this way. My claim here is that the ability to explain relationships and to recognize issues is partly dependent upon the language acquired in the earliest stages of primary socialization. This claim can be tested by listening to how most adults think about computers—that is, as a sender/receiver technology that is culturally neutral. The generations of cutting-edge thinkers in a variety of fields, including eminent scientists such as E. O. Wilson, who still think of organic processes as having the characteristics of a machine that can be re-engineered, is another example of how language both enables us to recognize and articulate relationships and, in the example of mechanistic thinking, how it can result in thinking being controlled and limited by the earlier ways of thinking encoded in the language. In the case of Wilson and other prominent scientists thinking of the brain as a machine can be traced back to beginnings of modern science in the sixteenth century. Other examples of primary socialization where the language carries forward earlier stereotyped thinking can be seen in how generations in the West were socialized to think that only men could be scientists, artists, historians, and theologians. As primary socialization was changed by giving students the language that named women who had made important contributions in these areas, it then became part of the student's taken-for-granted pattern of thinking.

As pointed out earlier, what is made explicit, what is passed on as taken for granted, what is based on cultural assumptions that are never examined, and what is presented in an abstract way that represents it as universally true, will influence the student's ability to think in ways that take account of the complexity of vital issues and to articulate a point of view that will lead others to reflect more deeply. The inability of most American students to recognize what was lost in terms of craft knowledge and community relationships when the assembly line was introduced is an example both of what has not been introduced by teachers in the process of primary socialization—and of how a restricted form of primary socialization limits the ability to resist traditions that undermine the economic well-being of communities. On the other hand, a process of primary socialization that provides the historical background, that includes a vocabulary that is as complex as the cultural and natural phenomena being presented to the students, that is related to their level of understanding and background experience, that makes explicit the underlying cultural assumptions (including alternative assumptions that would lead to recognizing new possibilities–or old ones that were mindlessly overturned), and that assesses the implications for the well-being of all the participants in the commons—will have a different affect on the students' communicative competence in the present and in the future. Hopefully, as adults their communicative competence will be used to address the unresolved eco-justice issues facing the world.

The key point here is that the teacher mediates between different approaches to the process of primary socialization—and this process of mediating between a restricted or an empowering educational experience is one of the constants at every grade level and in every subject area. The teacher's role as a cultural mediator even extends to the different moral values that are encoded in the language used to communicate in different subject areas ranging from the sciences, history, economics, literature—and teacher education. As most teachers have been socialized in the limited way that represents language as a conduit through which information, ideas, and data are passed, they are not likely to be aware that the language they use in different subject areas encodes the dominant culture's way of understanding relationships, the attributes of the participants in the relationships, and the culture's moral norms for how to act toward others (including the environment) that possess the culturally prescribed attributes. The language of colonization—of the environment and of other cultures—has only to be examined in order to see how language reproduces the moral norms of the culture.

An example that comes to mind is the way in which literacy has been represented in classrooms as an expression of progress, and "illiteracy" (that is, oral cultures) as backward. Ignorance about the complexity of oral cultures became the basis of a cultural prejudice which, in turn, became a moral judgment that opened the door for missionaries and other promoters of Western Enlightenment ideas. As the content of the curriculum can only be shared through the use of language, the teacher needs to understand that teaching one of the sciences, history, or whatever, is also a form of moral education (or, as is often the case, moral mis-education).

The proposal that students should construct their own knowledge, and that this will contribute to their realization of being autonomous individuals, ignores the inescapable reality of how the metaphorically layered nature of language carries forward past patterns of thinking that become in many instances the taken for granted basis of the student's thinking. The cultural patterns that the students take for granted also influence what they are able to reconstruct through critical thinking and problem solving, which is the source of another double bind. If the student, like most of us, does not question what is taken-for-granted, and the teacher has abdicated responsibility for including in the curriculum the aspects of culture that need to be understood as essential to the democratic process of identifying what needs to be changed and what needs to be conserved and renewed, we then have a form of education that leaves the students more vulnerable to the different expressions of industrial culture. While representing themselves as the promoters of freedom and autonomy, the silences in the constructivist teacher's approach to primary socialization furthers the forces that are undermining community and the natural environment.

⊠ Second Guiding Principle

The teacher's mediation between the different cultural forces and trends needs to contribute to the revitalization of the commons. Furthermore, the teacher must be knowledgeable about the differences in cultural ways of knowing and the characteristics of the local bioregion in order to avoid mediation's becoming a process of colonization.

While there are many forces contributing to the degradation of the environment (e.g., population pressures, destructive agricultural practices, exploiting fisheries beyond their ability to renew themselves, etc.) the industrial approach to production and consumption, which is predicated on the assumption that the "law of supply and demand" is the only safeguard needed for not exceeding the only natural limits, has the greatest adverse impact. The cultural assumptions underlying the spread of this industrial/profit oriented culture are reinforced in every facet of the public school and university curricula. In addition to the cultural assumptions about individualism, progress, evolution, an anthropocentric universe, mechanism as a model of thinking and problem solving, science as the most reliable source of knowledge, and the primacy of abstract ways of encoding knowledge in every area of the curriculum, students also encounter through the ever-present visual reminders of corporate logos and fast-food outlets on campuses that their futures and that of the corporate culture are inseparable—and that they will need to adjust their life goals accordingly.

Unlike the constructivist approaches where the content of the curriculum is largely decided either by the shifting interest of the students or by teachers who follow the central ideas of Dewey and Freire that education should be a process of "continuous transformation" (to quote Freire again), the teachers who are concerned with addressing eco-justice issues must not only be aware of the many ways in which the industrial/consumer-oriented culture are being reinforced in the classroom; they also need to be aware of the sustainable traditions that are part of the student's cultural background. That is, the teachers need to mediate in a way that helps the student understand the positive contributions of science and technology, as well as the aspects of the commons they undermine. In addressing the latter, the students need to examine the ways in which science and technology are still being used to further enclose (monetize) the commons, how they are beginning to be used to understand changes occurring in natural systems, and how to adapt human activity in ways that take account of the energy flows within natural systems. Students also need to examine the patterns of mutual support and other non-monetized activities and relationships within the different subcultures that are represented in the makeup of the classroom. A major focus of mediation should be on clarifying, in terms of everyday practices, the differences between industrial-oriented technologies and technologies that have been refined over generations of observing the changes occurring in the natural systems that are part of the commons. In light of these assumptions, the non-neutrality of the mediation process could not be clearer. Our future, in effect, depends upon students' learning to think ecologically. The changes in the environment are now dictating that we face up to this challenge, which means that this recommendation is based on environmental imperatives and not on an ideology.

The revitalization of the commons involves mediating between the dominant culture trend where excess—in the oversizing of SUVs, houses, meals, televisions, and body size—is being turned into a civic virtue and the still existing traditions of mutual support, mentoring, intergenerational knowledge of place, healing, and other activities that have not been turned into market opportunities. But mediation is not a matter of presenting students with an either/or set of possibilities and practices. Rather, it involves helping the students to recognize the trade-offs, how gains involve losses, that patterns of mutual support may involve the use of technologies and even industrially produced products, and that human activities need to be constantly assessed in terms of their adverse impact on natural systems—with the new commandment being "the less impact the better for the present and future generations."

Something more needs to be said about the teacher's role in mediating between the forces of modernization and traditions that are re-enacted in everyday life. As mentioned before, Western education is based on the Enlightenment assumption that the rational process reduces the need to rely upon traditions. Indeed, many prominent Western thinkers view traditions as an impediment to progress and in opposition to the various expressions of the rational process, including critical inquiry and experimental problem solving. As traditions include all the patterns, practices, technologies, ways of thinking, and so forth, that have been carried forward over four generations, the Western thinkers who view the rational process as a means of overcoming the hold of traditions do not recognize that their way of thinking of the rational process is itself a long-standing tradition.

Both Freirean and Deweyian thinkers view traditions as nonthinking habits and as impediments to the activities of "authentic subjects" engaging in the continuous renaming of the world. As I have pointed out earlier, this could not be a more traditional way of thinking—and one that has had a privileged role in the development of the industrial culture that has been trashing the environment for nearly two hundred years (a process that has escaped the attention of these supposedly rational thinkers). But the more important issue here is that in mediating between the various expressions of the culture of progress and the traditions that contribute to the renewal of the commons, the teacher needs to be clear that in helping students gain a more balanced and nuanced understanding of tradition, including the idea that taking seriously the importance of renewing traditions does not mean going back in time. Students also need to recognize that living traditions are not static, which they can understand better by examining the patterns of change and continuity within their own communities.

When traditions are understood in this way, educating for responsible citizenship in the commons is not a reactionary position. Nor should it be understood as recovering what has been lost. The challenge for the teacher is to help students identify the intergenerational knowledge and practices that have been marginalized by the attention being given to the traditions of modern, industrial culture that expand by undermining the non-monetized traditions of the commons. If the teacher can keep in focus that anything that has been passed along for over four generations or cohorts is a tradition, it will be easier for students to understand that the basic conflict is not between progress and tradition, between the industrial system of production and consumption and the intergenerational patterns of mutual support and non-monetized relationships. Rather, what needs to be clarified, in addition to the fact that not everything we do, think, or make becomes a tradition, is that traditions need to be assessed in terms of their impact on the commons and thus on the prospects for successfully addressing the eco-justice issues that confront the world's cultures. An example is the tradition of small farms (which has not entirely disappeared) that relies upon organic practices (again not a new tradition) that makes a constructive contribution to the commons—in terms of providing more nourishing food, economic interdependence, a sense of community, renewing the soil, providing the ground cover necessary for other species to exist, ensuring that the ground water is not contaminated with toxic chemicals, and so forth. The tradition of agribusiness, with its industrial approach to monocrops and use of pesticides and fertilizers that contaminate the ground water and become part of our body tissues, also needs to be assessed in terms of its impact on local communities and democratic decision making.

Other traditions that need to be assessed include the placing of supersized Wal-Mart-type stores on the outskirts of communities, and the tradition of small retailers located in the center of the town—where people interact in ways that are more likely to lead to conversations with a wider range of people and to discussions of topics that influence local democratic decision making. The tradition of megastores where everyone is expected to purchase tools that they may only use once or infrequently needs to be assessed in terms of the tradition of the library, which has been adapted in Berkeley, California, as a tool library where tools can be checked out when needed and returned for use by other people. *Renewing* the tradition of thinking of work as returned rather than as paid needs to be compared with the industrial approach to work. And again, the reference point for identifying which is more suitable is the impact that each has on the commons. I say, "renewed" because it is a tradition that is still taken for granted within different religiously based communities such as the Amish. It is practiced even among some families and neighbors in the dominant culture.

The above examples of cultural mediation highlight another point that is at the root of why the recommendations of the various constructivist theorists of learning contribute to the crisis of industrial culture. When teachers avoid framing the issues in terms of dichotomous categories (e.g. progress vs. tradition, consciousness raising vs. oppression, etc.), and avoid indoctrinating students to accept without questioning the Western god words of change, progress, autonomy, and emancipation, the students will be better informed about the importance of asking "What do we need to conserve?" As both Confucius and Wendell Berry point out, if we want to put right our relationships and our sense of moral accountability, we need to use language in a way that accurately expresses what we stand for, and what we take to be the fundamental relationships. This brings us to the third principle that should guide the teacher's responsibilities as a cultural mediator.

⬖ Third Guiding Principle

Mediating between the neoliberal ideology promoted in universities and by politicians and spokespersons for corporate culture and the conserving practices of environmentalists and others working to renew communities and cultures should be a constant focus of teachers who are concerned about eco-justice issues. While many cultures where constructivist-educational reforms are being introduced are based on religions, cosmologies, mythopoetic narratives (or a combination of all three) that emphasize the importance of looking to traditions as the source of authority, the teachers and students still are being exposed to the influence of the neoliberal ideology that underlies the values promoted in the Western media, consumer products, and in many of the policies of their own government. Thus, it is important for teachers in nonWestern classrooms to help students understand the broader implications of the ideological crosscurrents in their lives. Clarifying the ideological underpinnings of the ideas, values, interpretation of historical events, and even the metaphorical language in the different areas of the curriculum is also a central task of teachers in Western classrooms.

Suggesting that this should be an important focus is one thing. Expecting teachers trained in the constructivist ways of thinking to take it seriously moves into the domain of fantasy. It is important to keep in mind the nature of the assumptions constructivist teachers share (including Dewey and Freire) with the advocates of economic globalization. Even the constructivists who are critical of capitalism and globalization share with the more dysfunctional members of their extended ideological family the assumption that change is a constant—and the more of it the better. They also share the assumption (but for different reasons) that individuals must be emancipated from the community's network of traditions and responsibilities. Other shared assumptions include an anthropocentric way of thinking about human/nature relationships, the power of critical inquiry to free the individual or the group from the intergenerational knowledge of the community, an evolutionary way of thinking that represents the constructivist's one-best mode of thinking as more evolved than cultures that rely upon multiple ways of knowing. All the members of this dysfunctional family of liberal ideologues, while genuflecting once or twice in their writings in the direction of suggesting that some traditions should not be overturned, assume that traditions impede progress.

As the classroom is often located in communities where various non-monetized relationships and activities are still carried on, the teacher is in a situation where mediation is unavoidable. Silence on the teacher's part is even an expression of mediation, but it is one that will often leave the student caught between the two worlds of consumerism and intergenerational expectations of mutual support—between the West's illusion of a context-free form of individual freedom and self-determination and the constructivist's representation of the community as the source of the oppression. The Western media, curriculum materials designed for constructivist classrooms, and the spread of shopping malls that communicate the plenitude of the industrial way of life, and even the training of constructivist teachers—all reinforce the neoliberal ideology.

As a mediator between the ideology that promotes the spread of the consumer-oriented culture and what remains of the non-consumer practices and relationships within communities, teachers must take a position in opposition to the indoctrination and disempowerment that are hidden behind the liberal god words of individualism, autonomy, progress, and freedom. Regardless of the cultural setting, they can do this by asking the question that is never raised in the writings of the leading constructivist-learning theorists, or in the practice-oriented textbooks written by their followers. The question that should be asked in every area of the curriculum and at every grade level is: What needs to be conserved (and why), and what needs to be changed? That few teachers ask this question, which takes account of the cultural and natural ecology that constitutes the students' world, testifies to the degree of liberal indoctrination that occurs in the teacher's own education, and in the education of the people who write the curriculum materials. Asking these questions in the context of revitalizing the commons is very different from the misnamed conservatism that is being promoted by religious groups who ignore that what they take to be the word of God has been filtered through the thought processes of Biblical translators from different cultural backgrounds.

The irony is that while in their own lives teachers conserve (re-enact) the traditions they take for granted in their own homes and in the community, most teachers (and especially university professors) equate the words "conserve" and "conservatism" with a reactionary way of thinking. They also share the same basic misconception that is reinforced in the media when neoliberals are referred to as conservatives or neo-conservatives. The question What should be conserved (and why), and what should be changed? needs to be asked in classes in science, history, literature, political science, and in the use of computers and other technologies. Indeed, there is no area of the curriculum where the question should not be given extended attention.

In talking with others about why this question is especially important today, there is always the need to clarify how our political vocabulary is currently being misused by allowing neoliberals to masquerade as conservatives. This, in turn, requires explaining the contradiction in labeling environmentalists and people working to renew communities as liberals, especially since the industrial culture that is undermining both the environment and communities is based on the theories of classical liberal thinkers. Clarifying what conservatism stands for in terms of how we participate in the commons requires going back to the central ideas of Edmund Burke's conservatism. This leads to considering his two key ideas about ensuring that change makes a positive contribution to life in the community and that each generation has a responsibility for carrying forward the genuine achievements of previous generations so that future generations can build upon them. The next step in this process of clarification involves explaining why Burke's understanding of conservatism needs to be expanded in ways that take account of the natural ecologies that are the basis of the commons. Inevitably, the word "tradition" will come up in the discussion. The widespread misunderstanding of this word can only be addressed by connecting it with the everyday cultural practices, technologies, ways of thinking, and so forth, that go unnoticed as examples of tradition because they are a taken-for-granted part of experience. And when the conversation leads to understanding traditions in this more complex way, the question of which traditions should be changed and which conserved needs to be framed in terms of what contributes to the self-sufficiency of the community—which leads then to a discussion of the nature of the ecological crisis, and to how the traditions of industrial culture have contributed to it. Asking the question about what should be conserved is much like peeling an onion, where each layer that is peeled back leads to another hidden layer that, in turn, leads to a still deeper layer—until the core assumptions are reached.

To summarize the main point: the key question that needs to be asked in both Western and nonWestern classrooms does not fit what teachers in constructivist classrooms would consider to be politically correct. However, I suspect that in nonWestern classrooms many students will recognize the messianic and colonizing nature of the neoliberal ideology that underlies constructivist approaches to learning. And while their Western-educated teachers and educational bureaucrats may not recognize that the liberal assumptions underlying modernization are not universal principles that govern the "evolution" of all cultures, the parents and other members of the community still steeped in nonWestern traditions will be aware that the state's educational reforms are a source of intergenerational alienation. Whether the students will resist the pressures to become entirely dependent upon a money economy, which is promoted through the spread of industrial culture, is another question—especially when they realize that the promises of the industrial culture can only be achieved by the elites who control the system.

When I was first challenged to think about how teachers in the Quechua communities should teach the part of the curriculum prescribed by the government (sixty-five percent of which is to be based on Western content), I initially thought that I would present two different explanations of the teacher's mediating responsibilities. One explanation would address the pedagogical and curricular issues relating to eco-justice issues and the renewal of the commons in Western classrooms. The other explanation would then address the same issues, but within the context of nonWestern classrooms. Then I recognized that my lack of knowledge of the student's local culture meant that I had no right to make any recommendations about the teacher's mediating role in carrying forward and renewing the local traditions. This is the responsibility of the community in conjunction with the teacher, whether it be in a Zapotec, Muslim, or urban community in Brazil. The second realization was that when the curriculum is based on Western knowledge, the teacher's mediating responsibilities are the same for Western and nonWestern classrooms.

The latter statement may appear to contradict everything I have written before. However, if the Western part of the curriculum is understood not as representing the most advanced state of human knowledge and technological sophistication but as the conceptual basis of an individually and materialistically centered lifestyle that is being aggressively promoted around the world, then the suggestion that teachers, regardless of their cultural context, should view their role as cultural mediators in a similar way is not a contradiction. Regardless of whether the Western curriculum is being taught in Armenia, Uzbekistan, Thailand, Japan, New Zealand, or the United States, the same fundamental problems exist. That is, while these cultures vary widely, they all face an ecological crisis—which may take the form of toxic contamination of food, water shortages, loss of topsoil and ground cover, droughts caused by global warming, and so forth. In addition, they all face the loss of intergenerational knowledge as the Western media and its imitators in nonWestern countries promote consumerism and the other virtues of a Western lifestyle. They also face the economic hardships that result from a monetized economy, the industrial approach to production, and the economic competition that is promoted by the policies of the World Bank and WTO. In effect, the multiple dimensions of globalization now confront different cultures with a common set of challenges—which also include how to carry on the process of intergenerational renewal that is essential to maintaining a viable commons.

The dominant characteristic shared by all of the constructivist learning theorists, as well as their interpreters who control the content of the required teacher education courses, is that they all emphasize learning as a process. As noted earlier, socialization is also a process that enables one to think within the taken-for-granted patterns of the culture—and to become competent in the culture's languaging processes. But what distinguishes the constructivists' understanding of learning as a process of inquiry and the construction of new knowledge from the patterns of everyday socialization is that the constructivist view of process places the responsibility on the student for determining what is important to learn. Socialization in various social settings usually involves an adult, peer, or someone with the knowledge that is being shared who makes the initial decision about what is important to learn. For Dewey, the content of socialization—the ideas, issues, historical background account, etc.–cannot be specified in advance. Nor can the knowledge be taught in a systematic manner where knowledge is build upon a previously acquired foundation of knowledge. Rather, learning occurs for Dewey in the ongoing process of reconstructing experiences that are found to be problematic. That many taken-for-granted aspects of the student's cultural experience would not be recognized as problematic, did not occur to Dewey and his followers. The reason for Dewey's lack of awareness that what is taken for granted may represent a significant threat to the commons is that he lacked a theory of how language encodes earlier ways of thinking and how most of a person's culture is learned at a taken-for-granted level.

Piaget's emphasis on aligning the curriculum with the student's stage of cognitive development as well as his understanding of what constitutes the highest stage of reasoning and autonomy, also made any judgments about what a student needs to know both irrelevant and an expression of authoritarianism. Freire and his current followers are even more insistent that critical inquiry is the one and only valid approach to learning—in all areas of the curriculum. What Freire proposed in the *Pedagogy of the Oppressed* about the need for individuals, in order to achieve their highest potential as human beings, to rename the world and thus to become emancipated from the patterns of the previous generation is restated in his later writings. In *Pedagogy for Freedom* (1998), he writes that

> the educator with a democratic vision or posture cannot avoid in his teaching praxis insisting on the critical capacity, curiosity, and *autonomy* of the learner. . . . in the context of true learning, the learners will be engaged in a continuous transformation through which they become authentic subjects of the construction and reconstruction of what is being taught, side by side with the teacher who is equally subject to the same process. p. 33, italics added

The writings of his chief followers—Peter McLaren, Henry Giroux, Donaldo Macedo, and Peter Roberts—continue to restate what has now become a pedagogical formula. Namely, as transformative intellectuals, teachers are to guide students in reaching the highest level of process thinking, which is to subject all forms of knowledge and experience to critical inquiry. Roberts states the Freirean universal pedagogical truth that is to guide education in all the world's culture in the following way: "The ontological and historical vocation of *all* human beings is humanization," which means learning to think critically about how to transform existing structures—and to engage in the struggle of continuous liberation. (2000, p. 49–50; italics added), It would be hard to find a more direct statement of an educational theorist's imperialistic intentions. Unfortunately, this insight does not come easily to those who possess a messianic spirit, an ethnocentric understanding of the world, and the hubris that eliminates any doubt that they—and only they—know the truth that others should submit to.

As mentioned before, all of the constructivist-learning theorists, as well as a majority of their followers, ignore the ecological crisis and the differences in cultural ways of knowing. In light of the reforms that should be understood as related to the teacher's role as a cultural mediator, it needs to be restated that in sharing the core cultural assumptions that gave conceptual direction to the formation of the industrial culture of the West, all of the constructivist learning theorists should be viewed as reactionary thinkers. That is, their solution for addressing today's environmental problems is to impose the largely British tradition of classical liberal thinking—with its emphasis on change, individualism, the survival-of-the-fittest idea that emerges from critical reflection or economic practice, anthropocentrism, and viewing the rest of the world as culturally backward—on the rest of the world through the actions of classroom teachers who view their role as "transformative intellectuals."

⊠ Themes and Issues that Need to be Addressed in a Western Curriculum

Regardless if the classroom is in Syracuse, New York; Gallup, New Mexico; Cusco, Peru; Tokyo, Japan; Oxford, England; or any other city or village where the curriculum is based on Western knowledge, there are certain themes and issues that must be addressed. These cultural themes and issues are essential to understanding the cultural basis of the ecological crisis as well as the local cultural traditions that need to be revitalized in ways that can help reduce the rate and scale of environmental degradation. These themes and issues include: (1) the tensions between the sustainable practices within the commons and the spread of industrial culture—with its dual emphasis on consumerism and the further automation of production; (2) the difference between modern and indigenous (or low-impact) technologies; (3) the gains and losses connected with scientifically based technologies, including the colonizing nature of Western science; (4) the ideology that represents change as a progressive force and traditions as a source of backwardness—and the destructive impact of this view of tradition on revitalizing the commons and local democracy; and (5) the aspects of everyday life that can be enhanced by learning to think ecologically as opposed to thinking via an industrial model.

What is being recommended here thus requires a radically different approach to teacher education and, by extension, what is learned in the other university classes. Without an in-depth knowledge of their own and other cultures, teachers in Western countries will, in the name of individual autonomy and self-realization, reinforce the basic misconceptions that further undermine what remains of the community's traditions of self-reliance. In nonWestern cultures, the lack of in-depth knowledge of the cultural themes and issues listed above will likely result in the Western content of the curriculum's being presented as factual—and as the basis of the modern technologies and consumer goods that have such a seductive pull on the youth of many nonWestern cultures.

Given this potential lack of background knowledge, the following suggestions may help teachers mediate between the cultural differences implicit in the Western curriculum and the student's largely taken-for-granted culture. The first suggestion is to recognize that the curriculum, regardless of whether in a Western or nonWestern classroom, has a built-in status system. What is included in the curriculum is the outcome of a political process that represents a judgment about what is worth learning. What is excluded from the curriculum is also the result of a political process of decision making about what is unworthy—and thus has lower status. The political process of determining what is to be included and excluded may represent the efforts of the central government to undermine regional or ethnic differences and even to create a uniform workforce. It may also reflect deeply held biases, often reinforced by what is learned in universities, that the educated must share their supposedly more advanced knowledge with the less advanced segments of society.

The teacher who is mediating between different cultural forces (including economic, ideological, and ethnic differences) needs to treat the curriculum, regardless of content area, as representing *different* approaches to knowledge. Part of the process of cultural mediation involves helping the students to understand not only the nature of the status system, but also how the status system is dictated by the contradictions in the industrial culture. The local knowledge that is excluded from the curriculum needs to be made explicit and examined in terms of its contributions to sustaining the commons. Similarly, the high-status knowledge needs to be assessed in terms of its consequences for the commons. As students realize the importance of the knowledge excluded from the curriculum, as well as the cultural-transforming nature of high-status knowledge, they will be better able to make decisions that reflect what some scientists now refer to as the "precautionary principle." This principle, in short, states that "When an activity raises threats of harm to human health or the environment, precautionary measures should be taken even if some cause-and-effect relationships are not fully established scientifically"

(Rampton and Stauber, 2001, p. 134). Science, of course, is not the only basis for assessing the nature of a threat, which can also be directed at the well-being of the community. Housing projects, the building of a freeway, clearcutting the nearby forest, selling a small local enterprise to a transnational corporation—can all be threats that suggest the need for invoking the precautionary principle. There is also the possibility that the high-status knowledge encountered in the classroom may convey the message that the oral/intergenerationally based knowledge that is excluded is inferior—and that those who live by it are inferior.

When teaching Western science, the process of cultural mediation should include a comparison between the high-status mode of scientific inquiry and the local approach to observation, prediction, and control that have been practiced over generations. Examining the local approach to science, which may involve the guiding principle of moral reciprocity with the natural world and a long-term sense of responsibility for sustaining the commons for future generations, will bring out an important insight that can be marginalized by representing Western science as more advanced. The key concept that should be examined relates to thinking of the sciences, not in terms of one's being more advanced than the other, but as different sciences. And the examination of the differences needs to include the way in which local science is embedded in the mythopoetic/moral narratives of the culture—just as Western science is also influenced by the mythopoetic/moral narratives that have shaped Western cultures. That is, students need to understand the assumptions that underlie the Western approach to science: that it is based on the separation of facts and values, that it involves an objective way of knowing, and that the scientists are not responsible for how scientific knowledge is turned into technologies that may have unintended consequences.

In addition to learning about the method of inquiry and achievements of Western science, the dark side also needs to be examined. This includes considering the connections between science and the production of weapons of mass destruction; the increasing merging of science, technology, corporate culture, and universities; and the whole range of ethical questions that are connected with the genetic engineering of plants and animals. As part of the cultural comparison between Western and local science, the ways in which Western science undermines the moral values of nonWestern cultures also needs to be considered—including the connection between Western science and the consumer-dependent lifestyle that has become a major cause of environmental destruction. All of these aspects of Western science need to be assessed in terms of their impact on the life-supporting systems of the local commons—including their impact on the economic system that the local community depends on. Finally, mediating between the different approaches to science should lead to a consideration of which aspects of

Western science, and the technologies it has led to, can be adopted without undermining local traditions of self-sufficiency. In effect, learning about Western science in this way becomes a matter of exercising the "precautionary principle" where the new and supposedly superior approach to knowledge is carefully considered in terms of local traditions, needs, and patterns of moral reciprocity. The precautionary principle, when applied to the teacher's responsibility as a cultural mediator, leads to local democratic decision making that is not predetermined by the assumption that Western science is superior—and that not to embrace it in certain circumstances is a sign of cultural backwardness.

The treatment of technology, tradition, consumerism, as well as other themes and issues in a Western-oriented curriculum needs to be mediated in a similar way. Western technology is also represented in the curriculum as more advanced rather than as different from the technologies that have developed over generations of dwelling in a bioregion. Mediating between Western and local cultural approaches to technology should be focused on the fundamental question of which technologies contribute to the health of the local ecosystems and the systems of mutual support and self-sufficiency within the community. Further discussions need to address the ways in which Western technologies marginalize the development and use of craft knowledge, create a dependence upon a monetized economy, reduce the need for workers, undermine ceremonies as an integral part of certain forms of work such as planting and harvesting, transform work from a communal to a solitary experience, create dependence upon outside experts, and contribute to indebtedness through the constant need to purchase the steady stream of new technologies. As in the case of mediating between Western and local science, mediating between the local uses of technologies and the technologies being promoted as essential to economic development is not dependent upon taking university courses. Western universities are the seed beds for creating new technologies, but they offer few courses that provide students with the opportunity to examine the questions and issues cited above. And there are even fewer university courses where students learn about the nature and uses of technologies in non-Western cultures. The possibility that teacher education programs would provide guidance on how to address the questions about the differences between modern and traditional uses of technology that should be explored is nonexistent. Teacher education is based on the assumption that technology is both culturally neutral and the latest expression of progress. The history of how technology has influenced the area of teacher education has been one of blind acceptance and hypereuphoria about how the new technology would overcome previous limitations, followed by results that did not live up to initial expectations, followed by an unquestioning embrace of an even newer technology.

Teachers in both Western and nonWestern classrooms should not be reluctant to engage students in a discussion of the questions and issues I raised because they feel they have not taken a course in the subject area. If they rely upon their own as well as the students' knowledge of local traditions, whether in the areas of scientific explanations, uses of technology, or traditions that are relied upon, they will have a sound (indeed more accurate) basis for engaging students in a comparative discussion of cultural differences. After teachers have discussed with students the knowledge presented in the Western curriculum, they should identify the assumptions that Western knowledge is based upon—and then engage students in a discussion of how this knowledge has influenced developments in Western and nonWestern cultures. Again, teachers must trust their own ability to bring to the process of learning about the relationships and trends that they and their students observe in everyday life where Western influences are being intermixed with local traditions. A point that was made repeatedly in the earlier discussion of constructivist theories of learning and how they have affected teacher education program becomes critically important here. That is, the process approaches to classroom learning have totally marginalized the importance of teachers being knowledgeable about the nature of culture—including how its traditions are largely taken for granted and unconsciously passed on to students who will, in most instance continue to live by these taken-for-granted patterns. Teachers in both Western and nonWestern classrooms need to break from this tradition of ignoring the complex nature of cultural traditions and how they differ between cultures. In effect, teachers need to realize that they can compensate for this void in their own professional studies by becoming their own ethnographers—and helping the students to learn to observe the patterns they live by and to assess how these patterns (traditions) contribute to the health of the commons.

In addition to comparing the high-status content of a Western curriculum with local beliefs and practices, teachers also need to recognize that the process of cultural mediation is dependent upon understanding the languaging processes of the local culture as being profoundly different from the way language is represented in Western universities. As I do not have knowledge of how language is understood in nonWestern cultures, I will not make a generalization here. Therefore, the following comments should be understood in the context of the dominant way of thinking about language in the West, including the role that language plays in a Western-oriented curriculum. Again, the pedagogical implications need to be understood by teachers in both Western and nonWestern classrooms.

Western curriculum materials (and this is likely to be the case with curriculum materials written by educators in other countries who copy the Western patterns) are based on the assumption that language is neutral: that is, a conduit through which ideas and objectives, factual information can be passed. This view of language as culturally neutral, and as functioning as a conduit for passing ideas and information across from one person to others, is essential to maintaining the myth that there is such a thing as objective knowledge. It is also essential to the Western myth that is so central to constructivist-learning theorists, as well as to most academics. This myth represents the rational process, critical thinking, moral judgment, etc., as the activity of an individual—with the further assumption that the greater autonomy of the individual leads to a greater capacity to be rational and to think critically. The myth of language as a culturally neutral conduit for communicating "factual" knowledge and ideas is also essential to the process of cultural colonization, which is dependent upon the members of other cultures using one of the Western languages as the basis of their thinking and communicating. For example, the conduit view of language is essential to persuading the members of other cultures to accept such words as "development," "democracy," "individualism," "progress," "freedom," and "modernization" as representing possibilities that are universal in nature—and that are attainable as part of the process of cultural evolution.

The process of cultural mediation in Western and nonWestern classrooms needs to be based on an entirely different view of language. And this different view of language leads, in turn, to viewing intelligence in an entirely different way. When language, as I described it in an earlier chapter, is understood as encoding and carrying forward earlier culturally specific, metaphorically based, conceptual patterns of thinking derived from even earlier processes of metaphorical thinking, then intelligence can more correctly be understood as largely cultural in nature. That is, to restate the basic connection, as infants are socialized to the language that carries forward the conceptual schemata of their culture, their thinking and behavior is largely influenced by the taken-for-granted thought patterns encoded (for lack of a better word) in the language. That is, language thinks them as they think within the language. Even though language carries forward earlier patterns of metaphorical thinking, the individual, when facing a new situation, may recognize a different analogy as having more explanatory power. So the above statement about language doing the thinking (so to speak) has to be qualified so that it is understood as being a powerful and largely taken-for-granted influence—but not entirely deterministic.

The teacher who is mediating between cultural differences in Western and in nonWestern classrooms needs to be constantly aware of the metaphorical basis of what is represented in the curriculum as knowledge—about science, technology, history, literature, literacy, modernization, and so forth. As discussed in an earlier chapter, what is represented as high-status knowledge in the West is based on deep cultural assumptions or what can be called root metaphors. Examples of these root metaphors include patriarchy and anthropocentrism—the origins of which can be found in the mythopoetic narratives in the book of Genesis. These examples are important for another reason; namely, that the root metaphors that are the basis of other cultures are also found in their stories of creation—or what I am referring to as their mythopoetic narratives. Other root metaphors that influenced the development and are still the basis of further developments in the high-status knowledge of the West include mechanism, individualism, progress, economism, and now, evolution.

To summarize the role that root metaphors play in influencing the thought and behavior behind certain culture approaches to development: the root metaphors provide the interpretative framework for understanding relationships, how activities should be carried out, how problems are understood, the attributes possessed by the participants in the relationships—and thus the moral values that should govern the relationships. And the shaping influence of a root metaphor can be seen in many areas of the culture—and over a long period of time. The root metaphor also influences the process of analogic thinking where something new is being understood in terms of what is already familiar. Over time the analogy that prevails over other possibilities becomes encoded in iconic or image metaphors such as "data," "traditions," "development," "freedom," "creativity," "intelligence," and so forth. The early twentieth-century efforts to establish sociology as a science led to a process of thinking of sociological research as *being like* scientific research—and thus what was observed was also to be understood as having the same objectivity as what is yielded by Western science. To cite another example, the Western notion of creativity, if we trace the history of the metaphor, has shifted in meaning in ways that reflected the emergence of individualism as a root metaphor. And the influence of the root metaphor of mechanism can be seen in the process of analogic thinking that is going on in the fields of brain research, genetic engineering, medicine, education, agriculture, architecture, and so forth.

The challenge facing the teacher as a cultural mediator is in recognizing the taken-for-granted patterns of thinking that are present in the curriculum. And the problem is compounded by the fact that few if any teacher-education programs prepare teachers to become aware of the root metaphors, the political nature of analogic thinking, and the iconic metaphors that are present in every page of a textbook, piece of literature, scientific explanation, historical account—and present in every educational software program. What I am suggesting can and has been done by teachers who, after generations of not being aware, became aware of the root metaphor of patriarchy. Some teachers are becoming aware of anthropocentrism in curriculum materials. In both of these examples, teachers became aware of how language carried forward earlier culturally specific ways of thinking when social activists began to demand the end of domination—which was based in the languaging process that influenced every aspect of culture.

Since some teachers have proven that awareness of the shaping influence of language is possible, the next question becomes: What do the teachers as cultural mediators need to keep in mind in order to compensate for the silences in their professional studies? Again, I want to emphasize that the awareness of the influence of language on how the students learn to think about their culture and the influence on how they think about other cultures is an integral part of a cultural mediation pedagogy in both Western and nonWestern classrooms. If teachers in a Japanese, Bolivian, American, or other cultural classrooms are teaching part of a Western-based curriculum, they need to recognize that the words have a history and that they carry forward earlier patterns of metaphorical thinking that were influenced by the prevailing root metaphors of the culture. An example of how words encode an earlier, culturally specific way of thinking can be see in such terms as the "Middle East" and the "Far East." Even though I must travel west in order to reach Japan from my home in Oregon the conventional way of talking is to say that I am going to the Far East. The words encode the thought patterns of British colonialism, where London was the center of the world and to travel from London to Japan or Iraq was to go east. Similarly, the word "pioneer" is often used in American textbooks to refer to the settlement of the "wild" West. The metaphor carries with it the meaning of the first people in a new space—the "unsettled West" or in outer space. The word "pioneer" encoded the way of thinking of the Anglo and Euro-Americans who did not recognize the indigenous cultures that had settled the land for hundreds, even thousands of years as being human and thus as possessing any rights to the land.

In addition to helping students to recognize that words such as "development," "individualism," "literacy," "data," and so forth have a history and encode the assumptions that prevailed at an earlier time in a culture's history, teachers also need to clarify the nature of the root metaphors—including the historical circumstance that gave rise to them. When did thinking of organic life as machinelike come into the Western pattern of thinking? When did it represent a more useful way of thinking, and what useful technologies did it lead to? When has it led to ways of thinking and to the use of technologies that have had a destructive impact on the commons in Western and nonWestern cultures? How has this root metaphor reinforced other root metaphors such as "progress" and "evolution" that have become powerful ways of legitimating Western imperialism? How does the root metaphor of mechanism differ from the root metaphor of ecology that is now becoming the basic interpretative framework for environmentally conscious people? Is the idea of the commons unintelligible to a person whose thinking is based on a mechanistic root metaphor? Why is the person who thinks in terms of ecological design processes, culture as an ecology, and natural systems as ecologies, more receptive to thinking of the "commons" and to recognizing that the idea of the commons overcomes the traditional way of thinking of the human community as separate from the natural environment?

Mediating between different cultural patterns means helping students understand that language is not a conduit, but carries forward over many generations the earlier patterns of metaphorical thinking of a culture. That the metaphorical language of the high-status Western knowledge can also be a powerful source of cultural domination is also critical for all students to learn—both in Western and nonWestern classrooms. In Western classrooms, the domination may take the form of undermining or creating areas of cultural silence about what remains of the non-monetized intergenerational traditions of the community. The language may also carry forward and thus dominate the next generation in ways that preserve patterns of discrimination. And it may reinforce the development of technologies that are environmentally destructive while representing them as a further step in "progress" and in designing the natural systems we want (which may really be a matter of creating new market opportunities for transnational corporations). In nonWestern classrooms, comparisons between the local patterns of thinking, including the complex vocabularies that encode and carry forward the intergenerational stock of knowledge of the characteristics of the commons and the language in the Western-oriented curriculum needs to be continually brought out in ways that connect with the students' own language experiences in the community. Again it needs to be emphasized that the process of cultural mediation involves making comparisons between the abstract high-status knowledge represented in the curriculum, the cultural practices it leads to, and the local knowledge of the students and their community.

The moral, political, and ecological framework that needs to be kept clearly in focus is the long-term sustainability of the commons. Morally, this requires addressing the eco-justice issues of living in a way that does not damage the health of others and the prospects of future generations. Politically, the issues of local decision making also include the right to protect what contributes to the systems of mutual support and self-reliance that reduce the prospects of sinking into poverty and to destroying the environment in order to survive. Ecologically, it means helping students to recognize the ecological thinking that already exists in the community, how ecological thinking can lead to reliance on the sun rather than the current environmentally damaging sources of energy, and how the interdependent systems that make up the commons require thinking of the quality of life in terms of nurturing relationships rather than in terms of the acquisition of material goods.

In short, until revitalizing the commons and addressing eco-justice issues become the main foci of educational reform we will continue to be mislead by the false promises that are being promoted by both liberal/progressive educators and the transnational corporations that have as their shared goal a world monoculture of supposedly self-creating individuals who will be dependent upon consumerism.

references

Ackerman, Edith. 1996. "Perspective-Taking and Object Construction: Two Keys to Learning." In *Constructivism in Practice*. Edited by Yasmin Kafai and Michael Resnick. Mahwah, NJ: Lawrence Erlbaum.

Apffel-Marglin, Frederique, with PRATEC (editors). 1998. *The Spirit of Regeneration: Andean Culture Confronting Western Notions of Development*. London: Zed Books.

Armstrong, Jeannette. 1996. "'Sharing our Skin': Okanagan Community." In *The Case Against the Global Economy and for a Turn Toward the Local*. Edited by Jerry Mander and Edward Goldsmith. San Francisco: Sierra Club Books.

Barker, Debi, and Jerry Mander. 1999. *Invisible Government The World Trade Organization: Global Government for the New Millennium*. San Francisco: International Forum on Globalization.

Barlow, Maude. 2001. "Thirst for Justice." *Yes!* (Summer): 24–25.

Batalla, Guillermo Bonfil. 1996. *Mexico Profundo: Reclaiming a Civilization*. Austin: University of Texas Press.

Berger, Peter, and Thomas Luckmann. 1967. *The Social Construction of Reality*. New York: Doubleday Anchor.

Berry, Wendell. 2000. *Jayber Crow*. Washington, DC: Counterpoint.

———. 2000a. *Life is a Miracle: An Essay on a Modern Superstition*. Washington, D.C.: Counterpoint.

Bowers, C. A. 1997. *The Culture of Denial: Why the Environmental Movement Needs a Strategy for Reforming Universities and Public Schools*. Albany, NY: State University of New York Press.

———. 2001. *Educating for Eco-Justice and Community*. Athens: University of Georgia Press.

———. 2003a. "The Case Against John Dewey as an Environmental and Eco-Justice Philosopher." *Environmental Ethics*. 25 (Spring): 25–42.

———. 2003b. *Mindful Conservatism: Rethinking the Ideological and Educational Basis of an Ecologically Sustainable Future*. New York: Rowman & Littlefield.

Bowers, C. A. and Frederique Apffel-Marglin (editors). 2005. *Rethinking Freire: Globalization and the Environmental Crisis.* Mahwah, NJ: Lawrence Erlbaum.
Chaille, Christine, and Lory Britain. 1991. *The Young Child as Scientist: A Constructivist Approach to Early Childhood Science Education.* New York: Harper Collins.
Costilla, Karina. 2003. Unpublished field notes.
Dawkins, Richard. 1976. *The Selfish Gene.* New York: Oxford University Press.
DeVries, Rheta, and Betty Zan. 1994. *Moral Classrooms, Moral Children: Creating a Constructivist Atmosphere in Early Childhood Education.* New York: Teachers College Press.
DeVries, Rheta, Betty Zan, Carolyn Hildebrandt, Rebecca Edmiaston, and Christina Sales. 2002. *Developing Consructivist Early Childhood Curriculum: Practical Principles and Activities.* New York: Teachers College Press.
Dewey, John. 1916. *Democracy and Education.* New York: Macmillan.
———. 1935. *Liberalism and Social Action.* New York: Capricorn Books.
———. 1960. *Reconstructionist in Philosophy.* Boston: Beacon Press.
———. 1960. *The Quest for Certainty.* New York: G.P. Putnam's Sons.
Dewey, John, and Arthur F. Bentley. 1949. *Knowing and the Known.* Boston: Beacon.
Diamond, Jared. 2005. *Collapse: How Societies Choose to Fail or Succeed.* New York: Viking.
Doll, William E., Jr. 1993. *A Post-Modern Perspective on Curriculum.* New York: Teachers College Press.
Egan, Kieran. 1983. *Education and Psychology: Plato, Piaget, and Scientific Psychology.* New York: Teachers College Press.
———. 2002. *Getting It Wrong From the Beginning.* New Haven: Yale University Press.
Esteva, Gustavo, and Madhu Sari Prakash. 1998. *Escaping Education: Living and Learning Within Grassroots Culture.* London: Zed Books.
Freire, Paulo. 1973. *Education for Critical Consciousness.* New York: Seabury Press.
———. 1998. *Pedagogy of Freedom: Ethics, Democracy, and Civic Courage.* New York: Rowman & Littlefield.
———. 1974. *Pedagogy of the Oppressed.* New York: Seabury Press.
———. 1985. *The Politics of Education: Culture, Power, and Liberation.* South Hadley, MA: Bergin & Garvey.
Gadotti, Moacir. 2000. "Pedagogy of the Earth and Culture of Sustainability." Sao Paulo, Brazil: Instituto Paulo Freire.
Geertz, Clifford. 1973. *The Interpretation of Cultures.* New York: Basic Books.
Hausfather, Sam. 2003. "Where's the Content? The Role of Content in Constructivist Teacher Education." *Education: Annual Editions.* Edited by Fred Schultz. Guilford, CT: McGraw-Hill/Dushkin.
Ishizawa, Jorge, with Eduardo Grillo Fernandez. 2002. "Loving the World as It Is: Western Abstraction and Andean Nurturance." *ReVision.* (Vol. 24), No. 4: 21–26.
Jucker, Rolf. 2002. *Our Common Illiteracy: Education as if the Earth and People Mattered.* Frankfurt am Maim: Peter Lang.
Kelly, Kevin. 1994. *Out of Control: The Rise of Neo-Biological Civilization.* Reading, MA: Addison-Wesley.
Kohlberg, Lawrence. 1981. *The Philosophy of Moral Development: Moral Stages and the Idea of Justice.* New York: Harper & Row.
Lambert, Linda, Deborah Walker, Diane P. Zimmerman, Joanne E. Cooper, Morgan Dale Lambert, Mary E. Gardner, and Margaret Szabo 2002. *The Constructionist Leader.* New York: Teachers College Press.

Leopold, Aldo. 1949. *A Sand County Almanac*. San Francisco/New York: Sierra Club/Ballantine Books.
Light, Andrew, and Eric Katz. (editors). 1996. *Environmental Pragmatism*. London: Routledge.
McLaren, Peter, and Donna Houston. 1995. "The Nature of Amnesia: A Response to C. A. (Chet) Bowers." *Educational Studies* (Vol. 37) No. 2: pp. 196–205.
Norberg-Hodge, Helena. 1991. *Ancient Futures: Learning from Ladakh*. San Francisco: Sierra Club Books.
Norris, Kathleen. 1993. *Dakota: A Spiritual Geography*. Boston: Houghton Mifflin.
Oliver, Donald W., with Kathleen Gershman. 1989. *Education, Modernity, and Fractured Meaning: Toward a Process Theory of Teaching*. Albany, NY: State University of New York Press.
Oliver, Donald W., Jullie Canniff, and Jouni Korhonen. 2002. *The Primal, The Modern, and The Vital Center*. Brandon, VT: Foundation for Educational Renewal.
Orr, David. 2002. *The Nature of Design: Ecology, Culture, and Human Intention*. New York: Oxford University Press.
O'Sullivan, Edmund. 1999. *Transformative Learning: Educational Vision for the 21st Century*. Toronto: University of Toronto/Zed Books.
President's Committee of Advisors on Science and Technology, Panel on Educational Technology. 1997. *Report to the President on the Use of Technology to Strengthen K-12 Education in the United States*. Washington, DC.
Putnam, Robert. 1993. *Making Democracy Work: Civic Traditions in Modern Italy*. Princeton, NJ: Princeton University Press.
Rengifo, Grimaldo. 1998. "The *Ayllu*." In *The Spirit of Regeneration: Andean Culture Confronting Western Notions of Development*. Edited by Frederique Apffel-Marglin (with PRATEC). London: Zed Books.
———. 2001. *Children and Biodiversity in the Andes*. Alemania: Terre de Hommes.
———. 1998. "Education in the Modern West and in Andean Culture." In *The Spirit of Regeneration: Andean Culture Confronting Western Notions of Development*, edited by Frederique Apffel-Maglin (with PRATEC). London: Zed Books, pp. 172–192.
Roberts, Peter. 2000. *Education, Literacy, and Humanization: Exploring the Work of Paulo Freire*. Westport, CT: Bergin & Garvey.
Rorty, Richard. 1989. *Contingency, Irony, and Solidarity*. New York: Cambridge University Press.
Sale, Kirkpatrick. 1995. *Rebels Against the Future: The Luddites and Their War on the Industrial Revolution*. Reading. MA: Addison-Wesley.
Shils, Edward. 1981. *Tradition*. Chicago: University of Chicago Press.
Shiva, Vandana. 1990. "Development as a New Project of Western Patriarchy." In *Reweaving the World: The Emergence of Ecofeminism*. Edited by Irene Diamond and Gloria Feman Orenstein.
Shmulsky, Levi, Katie Marlowe, and Caitlin Daniel. 2003. "Quechua-Lamista Lifeways: The Digestion of Modern Educational Reforms." Unpublished manuscript.
Snyder, Gary. 1990. *The Practice of the Wild*. San Francisco: North Point Press.
Swimme, Brian. 1996. *The Hidden Heart of the Cosmos: Humanity and the New Story*. Maryknoll, NY: Orbis Books.
Swimme, Brian, and Thomas Berry. 1992. *The Universe Story*. San Fancisco: Harper San Francisco.
The Ecologist, 1993. *Whose Common Future? Reclaiming the Commons*. Philadelphia, PA: New Society Publishers.
Thurow, Lester. 1996. *The Future of Capitalism*. New York: W. Morrow.

Von Glasersfeld, Ernst. 1998. "Why Constructivism Must Be Radical." In *Cosntructivism and Education*. Edited by Marie Larochelle, Nadine Bednarz, and Jim Garrison. Cambridge, GB: Cambridge University Press.

Wilson, E. O. 1998. *Consilience: The Unity of Knowledge*. New York: Alfred A. Knopf.

Woolfolk, Anita. 1993. *Educational Psychology*. Boston: Allyn and Bacon.

WWW.EWCT.Com. pp. 2–5

index

A

About Schmidt (film), 74
Ackerman, Edith, 22
Albania, constructivist-learning theory in, 68–69
America
    achieving balance in rural areas, 59–61, 71–75
    constructivism in urban areas, 59–61, 71–73, 76–77
    hyperconsumerism versus eco-justice, 71–72
American Enterprise Institute, ix–x
analogic thinking, 45–48, 130
Ancient Futures (Norberg-Hodge), 100
Andean Project for Peasant Technologies (PRATEC), viii–ix
anthropocentrism as root metaphor, 49, 52, 54, 88, 129–31
anti-tradition tradition, 40–41
Apffel-Marglin, Frederique, 62
Arevalo, Mario, 63
Armstrong, Jeannette, 81
assumptions
    of industrial culture, 115–16
    and language, 131–32
    promoting consumerism via Western, 1–2
    subsistence lifestyle equals poverty, 60–61
    Western, derived from economic theories, 9
    See also constructivist assumptions
autonomy. *See* individual autonomy

B

Balinese, 59–61, 67–68
Barker, Debi, 89–90
Batalla, Guillermo Bonfil, 70–71
Bauhaus school of architecture, 50–51
Bentley, Arthur, 34
Berger, Peter, 35, 36–37
Berry, Wendell, 74
biblical examples of root metaphors, 49–50, 129–32

biodiversity
  constructivist ideology versus, 26, 88
  and cultural diversity, 11
  and culturally-sensitive approaches to education, 9–10
  Quechua's maintenance of, 62, 65, 67–68
  Western criticism of ecologically-centered cultures, 52
biological development
  intelligence as dependent on, 22, 53–54
  knowledge as result of, 36
  of mollusks, 21–22
  moral reasoning as dependent on, 23
  as readiness, 23–24
  Social Darwinism as basis of, 25, 26
  See also cognitive development of children
Bolivia, viii–ix, 91. See also Quechua and Aymara of the Peruvian Andes
Bruner, Jerome, 15
Burke, Edmund, 41
Canada, WTO and, 90
CATO Institute, ix
change
  assessing changes, 42, 119
  as automatic improvement, xi–xii, 16
  as core constructivist assumption, 4–5, 7, 17, 108, 119
  critical inquiry as path to, 13–14
  as normal condition of life, 16, 87–88
  as path to progress, xi, 2, 4–5, 6
  progressive linear form of, 19, 25, 26, 52, 53–54
  purpose of life versus, 95, 97
child-centered education. See constructivist-learning theories
children
  ancient future of, 70
  cultural-knowledge systems and learning process, viii, 62–63, 64
  decision-making process of, 57–58
  homeschooling, 69–70
  See also cognitive development of children
classical liberal ideology
  conservatives as, ix–x, xi, 40, 121
  and constructivists, 79–80, 124
  economic agenda of, 108–9
  promotion of consumerism, 90–91
  transnational corporations and organizations as, 87–88, 89
  See also constructivist-learning theorists; industrial culture; transnational corporations and organizations
Cochabamba, Bolivia, 91
cognitive development of children
  Dewey's scientific mode of inquiry, 18–19, 123
  Freire's levels of consciousness, 25–26
  Kohlberg's stages of moral development, 22–23
  Piaget's genetic epistemology and "readiness," viii, 21–23, 23–24, 62, 123
Collapse: How Societies Choose to Fail or Succeed (Diamond), 105
colonization of nonWestern countries
  assumption of prereflective "primitive" cultures, 25
  with constructivist pedagogies, 5, 95, 108
  with consumer-dependent lifestyle, xii, 1–3, 14, 19, 72, 76–77, 103–4
  contributory nature of university educations, 72–73
  cultural domination issues, 67–68
  Freirean-based social reforms as, 28, 123–24
  by industrial/market-oriented culture, xiii, 20
  as result of ignoring cultural context, 29
  versus traditions, 5, 8, 26
  via high-status knowledge concept, 2
  See also consumerism
commons
  assessing impact of traditions, 118
  autonomous individual versus, 1–2, 97
  decision making within, 81
  and democratic process, 81–82, 98
  enclosure of, ix–x, 80–81, 90–91, 116
  of indigenous peoples, 70–71
  individually constructed knowledge versus, 43–44
  industrial/consumer dependent culture replacing, xiii, 20
  intergenerational aspects of, 8, 23, 28, 32–34
  knowledge from ongoing dialogue with, 63

long-term sustainability, 77, 84, 87, 101, 125–26, 132–33
marginalization of, 6, 19
moral values as limiting, 18–19
as non-privatized way of stratifying wealth, 80
organized privatization of, 90–91
overview, 1, 5, 58–59
parenting within, 57–58, 58, 66–67, 69–70
in rural America, 73–75
"self" as relational, 81
self-sufficiency within, 19, 41–42, 71, 79, 93–94
teachers promoting revitalization of, 115–19
transforming into market opportunities, xi, 79–80
undermining of, 51
in urban America, 76–77
See also symbolic systems of the commons; traditions
communication
face-to-face versus digital, 83–84
and individual's cultural experiences, 36–37
language competence, 23, 37, 57, 58, 62–63, 113–15
metacommunication, 37, 110
See also language
community. *See* commons
computer-mediated learning, 7
computer technology, 7, 83–86, 98–99, 99
concrete operations stage of development, 22
conduit view of language, 16, 45, 84, 114–15, 132
conservatives. *See* classical liberal ideology
conserving orientation, 74–75, 88, 95, 118–19, 120–21
Constructivism in Practice (Ackerman), 22
constructivist assumptions
change equals progress, 4–5, 7, 17, 70–71
and critical inquiry, 38, 79–80
culturally destructive nature of, x–xi, 5–11, 87–88, 118–19
failure to examine, 7–8, 14–15
literacy equals developmental superiority, 2, 9, 115

oral cultures as primitive, 26, 115
superiority of constructed knowledge, 21, 115
superiority of consumerism, 1–2, 25, 60, 82
superiority of scientific method, 19
See also "knowledge cannot be transferred" dogma
constructivist-learning theories
commonalties, 26, 28–29, 43, 119
and computers in classrooms, 85–86
and corporate ideology, 92, 99
as imaginary culture, 70–71
learning as a process, 122–23
multicultural education paradox, 34, 39
nonWestern parents' response to, 66–67
pervasiveness of, viii–ix
as promotion of democratic decision making, 97–98
romantic view of self-creating individual, 96–97
as socializing students into consumerism, 3, 4, 87–88
student's interest guide curriculum, 77
Taiwanese abandonment of, 68
usefulness when applied correctly, 110
See also change; knowledge; "knowledge cannot be transferred" dogma; oppositional categories
constructivist-learning theorists
and classical liberal theorists', 79–80
and conduit view of language, 45
contradiction of theory and theorists, 53–54
Doll, 16, 31, 33
evidence of cultural transmission in, 33
inability to recognize personal biases, 54
liberal ideology of, 110
nonrecognition of cultural diversity, 67
nonsystematic methods, 98–99
See also Dewey, John; Freire, Paulo; Piaget, Jean
consumerism
classical liberal promotion of, 90–91
and constructivism, 3, 4, 87–88
consumption-dependent lifestyles, xii, 1–3, 19, 72, 76–77, 103–4
as cultural pattern, 38

debt as result of, 93–94, 127
versus eco-justice, 71–72
precautionary principle versus, 125–26, 127
reducing dependence on, 10–11
as result of Western influence, 54
revitalization of the commons versus, 116–17
Taiwanese promotion of, 68
and university educations, 73
Contingency, Irony, and Solidarity (Rorty), 95–96
corporations. *See* transnational corporations and organizations
critical inquiry/reflection
versus banking approach, 15–16, 27, 33, 64, 110
based on taken-for-granted assumptions, 38, 79–80
as center of process-oriented learning, 21
as central to student's construction of knowledge, 25
constructed knowledge as superior to known, 21, 115
constructivist definition of, 3
democratic versus subversive, 8
dual focus of, 13–14, 98–99
failure to assess assumptions, 14–15
versus heteronomy, 15–16
and Industrial Revolution, 79–80
language as impediment to, 34
origin of, 13
as path to change, 13–14
renaming the world, 4, 16, 27, 34, 96–97, 117
into social justice issues, 26–27
usefulness when applied correctly, x–xi, 41–42, 110
See also change
critically transitive consciousness, vii–viii, 25–26
critical pedagogy, 26–27
cultural domination issues, 67–68, 71–72, 132. *See also* change
cultural knowledge systems
devaluing, vii–viii, 18–19, 25–26
diversity of, 9, 15, 23, 34

environmental support within, 20
ignoring, viii, 22–23
influence of, 51–52
See also traditions
cultural/linguistic basis of learning, 15
cultural patterns
communicative competence and unawareness of, 37
and computers, 83–86
emphasizing explicit knowledge of negative aspects, 37–38
importance of, 29, 32–33
and language, 37
metaphorical thinking, 45–48
reproduction and renewal of culture, 34–35
root metaphors, 45, 48–54, 129–32
sharing of, 33, 37–38
and socialization, 122–23
versus socialization, 122–23
symbolic systems, 59, 72, 84
taken-for-granted nature of, 38–39, 43–46, 76–77, 107, 115, 130–31
transforming effect of Western ideas and values, 106
See also commons; language; traditions
cultural transmission model
as aspect of language processing, 32
essential cultural skills through, 63
as intergenerational renewal, 32–33
leading to heteronomy, 15
teaching all aspects of commons, 74–75
See also teachers as cultural mediators
curriculum
aligning with cognitive development stage, 21, 24–25
incorporating history of Western thought patterns, 131–32
science and technology considerations, 125–29
See also entries beginning with "teacher"

D

Dakota: A Spiritual Geography (Norris), 74
Darwinism, vii–viii, 22–23, 25, 26, 48, 90, 94.

See also cognitive development of children
Dawkins, Richard, 50
Death of a Salesman, The (Miller), 74
democracy and democratic decision-making
    and commons, 81–82, 98
    corporations as threat to, 91
    versus Dewey and Freire's approach to learning, x, xii, 42
    Dewey's interpretation of, 17, 19–20, 38, 42
    earth democracy, 81–82, 87, 92, 93
    education leading to, 19
    marginalization of, 4
    moral reciprocity with non-human world, 81–82
    for overturning bad traditions, 41
    and paradox of constructivist-learning theories, 38, 97–98
    teaching about, 111–12
    traditions as anathema to, 8, 19, 27–28
    value determination within, 100
Dewey, John
    as advocate for participatory decision making, 25, 96, 97
    contradictory interpretation of democracy, 17, 19–20, 38, 42
    Darwinian evolution as rationale for, 53–54
    experimental inquiry approach to knowledge, viii, x–xi, 18–19, 34, 87–88, 110
    failure to recognize cultural diversity, vii, x, 17, 19
    failure to value cultural traditions, xi, 17–18, 123
    Freire compared to, 29
    intelligence as reconstructing experience, 15, 18
    on language interfering with experimental inquiry, 34
    overview, 3–4, 17–20
    Piaget compared to, 25
    purpose of education, 18
Diamond, Jared, 105
diversity. *See* biodiversity
Doll, William, Jr., 16, 31, 33

E

earth democracy, 81–82, 87, 92, 93
eco-justice
    of commons versus corporations and organizations, 81
    culturally-informed approach to education, 104–6
    hyperconsumerist lifestyle versus, 71–72
    teaching about issues of, 116
ecological crises
    and constructive approach to education, 29, 78
    Dewey's silence about, 17–18
    Freire's dismissal of, 107
    and poverty, 36
    as result of consumption-based economy, 5, 71–72, 77
    from teachers' education in constructivist theory, 10–11
    teachers' position on, 112
    as worldwide commonality, 122–23
    See also consumerism
ecological footprints, 67–68, 73, 101, 105–6
ecological sustainability
    examining technology's role in, 127–28
    versus mechanization, 52
    teacher awareness of systems for, 35
    See also sustainable futures
ecology of natural systems
    versus anthropocentrism and progress, 52–53
    conserving orientation, 74–75, 88, 95, 118–19, 120–21
    conserving orientation in rural America, 74–75
    constructivist unconcern with, 20
    corporate unconcern with, 95
    versus evolution, 49
    focusing on industrial-oriented versus natural systems, 116
    importance of, 32
    marginalizing with mechanistic root metaphor, 51–52
    Quechua adjustments to, 62

survival-of-the-fittest versus, 2, 5, 90, 124
utilizing for basic needs, 75
economic level playing field theory, 90, 100
Educating for Eco-Justice and Community (Bower), 110
Educational Psychology (Woolfolk), 50
educational reform
    computers and constructivist, 85–86
    culturally-informed eco-justice approach, 104–6
    culturally-sensitive approaches to, 9–10
    guidelines for, 105–6
    imposition of Western model of development, 26
    maintaining importance of basic relationships, 103–4
    poverty from, viii–ix
    for a sustainable future, 103–4
    See also teachers as cultural mediators
Education for Critical Consciousness (Freire), vii–viii, 25–26
Egan, Keiran, 21–22, 25
enclosure of the commons, ix–x, 80–85, 90–91, 116
England, Statute of Merton (1235), 80
Enlightenment, the
    as colonizing mindset, 28
    as progressive way of thinking, 87
    and Smith's invisible hands, 66
    traditions as oppressive, 19, 39–40, 117
environment. *See entries beginning with* "eco"
Escaping Education (Esteva and Prakash), 69–70
Esteva, Gustavo, 28, 69–70
ethnocentrism. *See* Western ethnocentrism
evolution
    as basis of Western superiority complex, vii–viii
    versus ecology, 49
    Social Darwinism, vii–viii, 22–23, 25, 26, 48, 90, 94
    as worldwide metanarrative, 109
experimental inquiry, viii, x–xi, 18–19, 34, 87–88, 110

F

formal operations stage of development, 22
free-market system, ix–x, 79, 89–90. *See also* consumerism
Freire, Paulo
    "banking" approach to education, 15–16, 27, 33, 64, 110
    Darwinian evolution as rationale for, 53–54
    on decision-making by children, 57–58
    on democratic teaching, 27–28
    Dewey compared to, 29
    ignoring cultural differences, 109
    overview, 4, 25–28
    Piaget compared to, 28–29
    stages of cognitive development, vii–viii, 25–26
    on true words and action reflection, 27
    unawareness of mutually supportive traditions, 94–95, 107
Freirean tradition, 26–27

G

Gardner, Howard, 24
Geertz, Clifford, 86–87
genetic engineering, 51
genetic epistemology, viii, 21–23, 23–24, 62, 123
genetics, determining intelligence with, 48
Glaserfeld, Ernst von, 21
globalization
    Industrial Revolution as center of, 9
    replacing intergenerational knowledge with product awareness, 79–80
    resistance to, 34–35, 91–92, 93
    television, films, and, 2
    of Western consumerist lifestyle, xii, 1–3, 14, 19, 72, 76–77, 103–4
    See also consumerism; transnational corporations and organizations

H

Hausfather, Sam, 107
hegemony. *See* colonization of nonWestern countries
Heidegger, Martin, 44
high-status knowledge
    addressing eco-justice issues with, 72
    from American universities, 71–72
    assessing consequences for commons, 125–26
    change equals progress, 4
    cultural domination aspects of, 132
    versus low-status knowledge, 2
    marginalization of traditions, 4–5, 51–52
    as outcome of individually-centered rational process, 7–8
    and root metaphors, 51
    as superior to nonWestern traditions, 26
Hobbes, Thomas, 50
Hoover Institute, ix–x
Hores, Carlos Ortega, 62–63
hyperconsumerism. *See* consumerism

I

iconic metaphors, 45–48
ideal liberal society, 96
ideology, defining, 86–87
image words, 45–48
imaginary culture, constructivist-learning theory as, 70–71
imperialism. *See* colonization of nonWestern countries
India, 91–92
indigenous peoples
    advantages of commons culture, 70–71
    in America, 73
    colonizing with Freirean ideas, 28
    earth democracy practice, 81–82
    environmentalism (or not) of, 109–10
    resistance to industrial model, 92, 93
    and Western teachers, 106
    Zapotecs of Mexico, 69–70
    See also Quechua and Aymara of the Peruvian Andes
individual autonomy
    aligning curriculum with cognitive development stage leads to, 21, 24–25
    anthropocentrist root of, 54
    versus commons of indigenous culture, 71
    computer-enforcement of, 84–85
    Freire's vision of, 123–24
    versus heteronomy, 15–17, 22–23, 24–25
    as ideological construction, 28
    versus intergenerational connection, 1–2, 83–84
    as lonely and exhausted by effort, 97
    versus recognizing cultural patterns, 115
    as state of continual becoming, 88
    as superior adaptation to changing world, 17
    value determination by, 100
    work/debt cycle as aspect of, 93–94
industrial culture
    versus community self-sufficiency, 79, 93–94
    crises as result of, 76–77
    cultural assumptions underlying, 115–16
    environmental devastation from, ix–x, 20, 77
    globalism of, 5
    machine model, 50–51
    marginalization of workers' craft skills, 82
    meaning of life in, 95
    profit motive, xi
    resistance to, 34–35, 91–92, 93
    as self-consuming, 5
    status system embedded in, 125–26
    teaching children how to resist, 115
    and Western culture, 1–2
Industrial Revolution, 9, 79
intelligence
    as ability to adapt to changes in environment, 22
    as cultural, 45
    cultural aspect of, 45–46
    English language test as measure of, 48
    genetic determinations of, 48
    logico-mathematical stage of development as, 22

objective measures of, 47–48
as ongoing process of reconstructing experience, 15, 18
scientific mode of inquiry equated to, 22, 43
intergenerational knowledge
as basis of self-identity, 83–84
constructivist view of, 119
versus individual autonomy, 1–2
perpetuation of commons with, 8, 23, 28, 32–34
replacing with product awareness, 79–80
and standards of social justice, 92–93
teacher awareness of, 33–38, 114
Western bias against, 60
See also traditions
intergenerational renewal
of commons, 59
cultural transmission model as, 32–33
and sustainable futures, 77, 84, 87
See also cultural transmission model
intergenerational traditions. *See* traditions
Islamic lifestyle and constructivist-learning theory, 68–69, 92

J

Japan, 68, 90
Jayber Crow (Berry), 74
Jucker, Rolf, 72, 103

K

Kepler, Johannes, 50
Knowing and the Known (Dewey and Bentley), 34
knowledge
Dewey's spectator approach to, vii, 19–20, 33, 57, 64
discounting traditions as inferior, 26
from experimental inquiry, viii, x–xi, 18–19, 34, 87–88, 110
Freire's "banking approach" to education, 15–16, 27, 33, 64, 110
in head and hands, eyes, nose, soul, 63
high-status versus low-status, 2
objective knowledge, 16, 45, 128–29
Piaget's autonomy versus heteronomy, 15–17, 22–23, 24–25
shared cultural, 33, 37–38
See also cultural knowledge systems; high-status knowledge; intergenerational knowledge
"knowledge cannot be transferred" dogma
as abstract thinking, 40
versus children's ancient future, 70
versus cross-culturally informed interpretation of learning, 15–16
Freire's defense of, 27, 57–58
Glaserfeld's explanation of, 21
and language, 15–16, 21, 23, 27, 40
versus language and cultural patterns, 115
and Piaget's view of language, 23
knowledge systems. *See* cultural knowledge systems; intergenerational knowledge
Kohlberg, Lawrence, 22–23

L

labor mediators versus teachers as mediators, 111, 112–13
language
communicative competence, 23, 37, 57, 58, 62–63, 113–15
conduit view of, 16, 45, 84, 114–15, 132
as core of culture, 37, 128
and cultural transmission learning model, 32–33, 35–36
enabling and controlling aspects of, 113–14
historical assumptions tied to, 131–32
image words, 45–48
"knowledge cannot be transmitted" via, 15–16, 21, 23, 27, 40
linguistically encoded language, 45–46
mediating role of, 15, 132
and metaphorical thinking, 45–48
and moral norms of culture, 115
as post-cognitive structure development, 23

as repressive, 34
  role in influencing awareness, 34, 35–36, 49–50
  root metaphors, 45, 48–54, 129–32
  and thinking patterns, 45, 49–50
  as tool for assessing traditions, 4–5
  as useful cultural knowledge, 32–33
  and Western curriculum materials, 128–29
laws of natural selection as basis for cultural development, 2, 5, 90, 124
learning, multiple methods of, 15
learning theories
  requirements for, 39–40
  thought processes of theorists, 53–54
  See also constructivist-learning theories; cultural transmission model
Leopold, Aldo, 17
Leviathan (Hobbes), 50
Liberalism and Social Action (Dewey), 42
literacy
  and computers, 83
  as evidence of developmental superiority, 2, 9, 115
  Freire's followers' loss of faith in, 28
  teaching ecological literacy, 103–4, 131
literacy, Western view of, 2
logico-mathematical stage of intelligence, 22
Luckmann, Thomas, 35, 36–37

M

Making Democracy Work (Putnam), 93
Mander, Jerry, 89–90
market relationships within the commons, 80
Marx, Karl, 86
mechanistic root metaphors, 51–52
mediating and mediators
  computers as, 83–84
  in labor disputes, 111, 112–13
  role of language, 15, 132
  silence as expression of mediation, 120
  See also teachers as cultural mediators
metacommunication, 37, 110
metaphorical thinking
  cultural specificity of, 51
  iconic metaphors, 45–48
  root metaphors, 45, 48–54, 129–32
Mexico, constructivist-learning theory in, 69–70
Mexico, NAFTA's effect on, 8–9
Milestone (Qutb), 92
Miller, Arthur, 74
Minsky, Marvin, 50
modern liberal theorists, 50
monetized economy. *See* consumerism
Monsanto, 91
moral reciprocity
  and face-to-face communication, 83
  indigenous peoples awareness of, 81–82
  and nonWestern approach to science, 126
  in rural America, 74
  as transmitted knowledge, 93
  versus Western assumptions, 1–2
moral values
  applying scientific method to, 18–19
  democratic determination of, 100
  developmental stages for, 22–23
  versus genetic engineering, 51
  and language, 115
  and root metaphors, 52–53
multicultural education, 34, 39
Muslims and constructivist-learning theory, 68–69
mutual support systems
  enclosure of, 72, 80–85, 100
  Freire ignorance of, 94–95, 107
  as nonWestern option to autonomy, 67
  in rural America, 73–74
  Western disbelief in, xi, 1, 60–61
  See also commons
mythopoetic narratives, 23, 40, 83–84, 95, 126. See also symbolic systems of the commons

N

NAFTA, poverty in Mexico from, 8–9
natural ecologies. *See* ecology of natural systems
natural selection, 2, 26, 33. *See also* Darwinism

nature, learning to read signs of, 63–64, 65. *See also* ecological sustainability
Nature of Design, The (Orr), 103–4
neo-liberal ideology, 90–91, 112–13, 119–24
Nietzsche, Friedrich, 46, 49
Nigeria, 91
Norberg-Hodge, Helena, 100
Norris, Kathleen, 74

O

Oba, Jorge Ishizawa, 63
objective knowledge, 16, 45, 128–29. *See also* "knowledge cannot be transferred" dogma
Ojibway Nation (Ontario, Canada), 81
Oliver, Donald, 16
"open system" versus "closed system," 16, 31, 33
oppositional categories
    autonomy versus heteronomy, 15–17, 22–23, 24–25
    critical reflection versus banking approach, 15–16, 27, 33, 64, 110
    experimental inquiry, viii, x–xi, 18–19, 34, 87–88, 110
    experimental inquiry versus spectator approach, vii, 19–20, 33, 57, 64
    "open system" versus "closed system," 16, 31, 33
    See also critical inquiry/reflection
oral cultures, 26, 115
Orr, David, 103–4
Our Common Illiteracy: Education as if the Earth and People Mattered (Jucker), 72, 103

P

Pakistan, constructivist-learning theory in, 68–69
parenting, 57–58, 58, 66–67, 69–70
participatory decision making, 25, 96, 97
patriarchy as root metaphor, 49, 51–52, 54, 129–31
Pedagogy of Freedom (Freire), 27–28, 123

Pedagogy of the Oppressed (Freire), 16, 27, 123
Peru. *See* Quechua and Aymara of the Peruvian Andes
Piaget, Jean
    autonomy versus heteronomy, 15–17, 22–23, 24–25
    cognitive development scheme of, viii, 21–23, 23–24, 62, 123
    Darwinian evolution as rationale for, 53–54
    Dewey compared to, 25
    Freire compared to, 28–29
    genetic epistemology and "readiness," viii, 21–23, 23–24, 62, 123
    mollusk-related insights of, 21–22
    overview, 20–25
    See also "knowledge cannot be transferred" dogma
political role of metaphoric thinking, 51
politics, teachers' role in explaining, 111
poverty
    from educational reforms, viii–ix
    as outcome of consumer-based economy, 71, 76–77, 100, 103, 105
    as result of destroying commons, 8–9, 36, 94, 105
    subsistence lifestyle versus, 60–61
Practice of the Wild, The (Snyder), 82
Prakash, Madhu Suri, 69–70
PRATEC (Andean Project for Peasant Technologies), viii–ix
precautionary principle, 125–26, 127
preoperational stage of development, 22
President's Committee of Advisors on Science and Technology, 85–86
primary socialization, 113–15
privatization of the commons, ix–x, 80–85, 90–91
progress
    assessing the worth of, 41–42
    change equated to, 2, 4–5
    constructivist-based education and Western myth of, 29
    and god words of modernization, x, 6, 26, 80, 118–19, 120
    as higher values than religious beliefs, 100–101

linear form of, 19, 25, 26, 52
versus moral values, 51, 52–53
See also change
public schools and universities, Western, 2, 19, 71–73, 106–7, 127
Putnam, Robert, 93

Q

Quechua and Aymara of the Peruvian Andes
    analyzing effect of constructivism on commons of, 59–61
    children's cultural way of knowing, 62–63, 64
    communication with natural world, 63–64, 65
    constructivist ideas versus cosmovision of, 64–65
    ecological accomplishments of, 61–62
    effect of constructivism on commons, 65–67
    potential effect of constructivism, viii–ix
    teacher training and agenda, 61
Quest for Certainty, The (Dewey), 18
Qutb, Sayyid, 92

R

rational thought
    and autonomy, 129
    and cultural influence, 45, 48–50
    knowledge as outcome of, 7–8
    tradition as block to, 39
    and traditions, 117–18
    Western science as highest expression of, 2
    See also critical inquiry/reflection
reactionary thinking, asking what to conserve as, 4, 120–21
readiness, 23–24
Reading and Writing for Critical Thinking Project, 3
Rebels Against the Future (Sale), 79
Reconstruction in Philosophy (Dewey), vii
Reddy, Michael, 45

religious beliefs and constructivism, 68–69, 92, 99–100
renaming the world, 4, 16, 27, 34, 96–97, 117
Rengifo Vasquez, Grimaldo, 28, 64–65
Report to the President on the Use of Technology to Strengthen K-12 Education (President's Committee of Advisors on Science and Technology), 85–86
Roberts, Peter, 123–24
root metaphors, 45, 48–54, 129–32
Rorty, Richard, 95–96
Rousseau, 14
rural areas, commons in, 73–75

S

Sale, Kirkpatrick, 79, 94
Sanchez, Loyda, 28
science
    curriculum considerations, 126–27
    as high-status knowledge, 2
    limited boundaries of, 23
    mechanistic root metaphors, 51–52
    mediating positive and negative aspects of, 10–11, 116
    scientific mode of inquiry, 18–19
    as superior to commons, xi
self-sufficiency, tradition of, 19, 41–42, 71, 94
semi-intransitivity of consciousness, vii–viii, 25, 26
sensori-motor stage of development, 22
shared knowledge, 33, 37–38
Shil, Edward, 39, 40–41
Shiva, Vandana, 60, 61
Siddhartha, 28
Silver, Lee, 51
Snyder, Gary, 82
Social Construction of Reality, The (Berger and Luckmann), 37
Social Darwinism, vii–viii, 22–23, 25, 26, 48, 90, 94
socialization
    Dewey's view of, 123
    language considerations, 47
    in Rorty's ideal liberal society, 96

and taken-for-granted cultural patterns, 122–23
teacher as mediator of primary socialization, 113–15
and use of computers, 86
social justice issues, critical reflection on, 26–27
spectator approach to knowledge, vii, 19–20, 33, 57, 64
Spirit of Regeneration, The (Apffel-Marglin, ed.), 62
Statute of Merton (1235), 80–81
student-centered learning. *See* constructivist-learning theories
subsistence cultures, 61
survival-of-the-fittest versus environment, 2, 5, 90, 124
sustainable futures
    democratic decision-making within commons, 81–82
    intergenerational renewal, 77, 84, 87
    recommendations for, 103–4
    specificity of, 109
    teachers as cultural mediators for, 106–7, 116, 125–26, 132
    transdisciplinary approach to, 72, 74–75
    WTO and World Bank threat to, 88–91
    See also commons; ecological sustainability
symbolic interactionism, 34
symbolic systems of the commons
    and biodiversity, 65
    constructivist ignorance about, 77
    enclosure of, 72, 84, 85, 100
    in general, 59
    helping students connect with, 73–74
    mythopoetic narratives, 23, 40, 83–84, 95, 126
    non-constructed nature of, 70
    teacher education in, 124–33
    in urban America, 76
    See also traditions

T

Taiwanese, 68
teacher awareness of local culture
    constructivist programs versus, 106–8
    criticisms of, 108–10
    as ecologically desirable, 10–11
    and history of Western thought patterns, 131–32
    importance of, 104–6
    language considerations, 128–29
    mediating eco-justice issues and corporate culture, 119–24
    nature of, 125
    promoting revitalization of the commons, 115–19
    science and technology considerations, 125–28
    socialization and cultural patterns, 122–23
    understanding intergenerational knowledge, 33–38, 114
    and Western ethnocentrism, 106–7
    See also teachers as cultural mediators
teacher education in constructivist theory
    constructivist approach to, 2–4, 6
    failure to clarify cultural assumptions, 7–8
    judging student's stage of cognitive development, 21, 23–24
    "knowledge cannot be transmitted" dogma, 15–16
    method for creating "purified medium of action," 19
teaching moral reasoning, 23
teaching Western curriculum in nonWestern classrooms, 122–23
textbooks for, 14–15, 20–21, 25
teachers
    as catalysts triggering self-organization, 16
    curriculum, 21, 24–25, 125–29, 131–32
    existential choices facing, 105
    importance of taking a stand, 112
    issues with determining "readiness," 24
    in rural America, 74
    as transformative intellectuals, x, 93, 98–99, 107, 123–24
    of Western curriculum in nonWestern classrooms, 122, 124–33
    See also teacher's moral responsibilities
teachers as cultural mediators
    within commons, 45

between corporate culture and eco-justice issues, 119–24, 131–32
between cultural orientations, 106–7, 132
curriculum considerations, 125–28
effect of constructivist training on, 110–11
guidelines for, 124–33
overview, 111–13
process of primary socialization, 113–15
promoting revitalization of the commons, 115–19
of quality-of-life issues, xiii, 10–11
recognizing taken-for-granted patterns of thinking, 130–31
rural versus corporate or urban lifestyle, 74–75
for sustainable futures, 106–7, 116, 125–26, 132
and symbolic systems of the commons, 73–74
between traditions and forces of change, 108
understanding intergenerational knowledge, 33–38, 114
teacher's moral responsibilities
addressing eco-justice issues, 72
clarifying ideological underpinnings of ideas, 119
differentiating between monetized and non-monetized aspects of the commons, 77, 116
examining root metaphors, 51–53
protecting traditions, xii, 43
teaching about tradeoffs, 117
understanding traditions, 33–35, 41–42
technology
assessing the worth of, 41–42
cultural non-neutrality of, 44
curriculum considerations, 127–28
as high-status knowledge, 2
as means for subjugating nature, 82
mediating positive and negative aspects of, 10–11, 116
replacing self-sufficiency with, 5, 19
shaping influence of, 107
terminator seed, 91
textbooks on constructivist theory, 14–15, 20–21, 25, 128–29

theories of learning
requirements for, 39–40
thought processes of theorists, 53–54
See also constructivist-learning theories; cultural transmission model
Thurow, Lester, 47
traditionalisms, 41
traditions
attempts to retain, 6
consumerism as replacement for, 2
cultural associations to, 39
cultural complexity of, 96
deciding whether to abandon or keep, 109, 117–18
as externally controlled nonreconstructed experience, 15, 18
as final vocabulary, static, 96
fluid nature of, 41, 117–18
forces of modernization versus, 117
four generations definition, 40
globalization versus, 88
inclusive of daily taken-for-granted realities, 40–41
inconsequential nature of, 95
marginalization and subversion of, xi–xii, 4–5, 18, 51–52
root metaphors as, 50
self-sufficiency, 19, 41–42, 71, 94
as source of backwardness, 14
as source of empowerment, 35
taken-for-granted nature of, 38–39, 43–44, 45–46
thinking of language as unbiased conduit, 45
understanding, 33–35, 39–44
Western colonization versus, 5, 8, 26
See also cultural patterns; language
Tradition (Shil), 39
transformative intellectuals, x, 93, 98–99, 107, 123–24
transformative learning. *See* constructivist-learning theories
transitivity of consciousness, vii–viii, 25, 26
transnational corporations and organizations
approach to agriculture, 75
as classical liberal thinkers and theorists, 87–88, 89

destructive power of, 100
dominance in America, 71–72
and eco-justice issues, 119–24
versus eco-justice of commons, 81
exploitation of constructivist ideas, 4
globalizing Western consumerist lifestyle, 14
paradox of selling goods and displacing workers, 5, 8–9, 94, 106–7
World Bank, 72–73, 87–88, 90–91, 100
World Trade Organization, 88–90, 96, 101, 122
Turkey, constructivist-learning theory in, 68–69

U

unholy trinity, 105
universalists, 88. *See also* colonization of nonWestern countries
universities, Western, 2, 19, 71–73, 106–7, 127
urban areas, commons in, 76–77

V

Vygotsky, Lev Semyonovich, 15, 34, 35

W

Wabigoon Lake Ojibway Nation (Ontario, Canada), 81
Wal-Mart, 91, 118
Western cultures
  anthropocentrism as root metaphor, 49, 52, 54, 129–31
  bias against intergenerational knowledge, 60
  and computers, 85–86
  cultural domination issues, 67–68, 71–72, 132
  history of inequities, 104–5
  industrial process as tantamount, 1–2
  low status of cultural and environmental commons, 71–72
  mutual support systems as fantasy to, xi, 1, 60–61
  orientation of public schools and universities, 2, 19, 71–73, 106–7, 127
  patriarchy as root metaphor, 49, 51–52, 54, 129–31
  subsistence lifestyle as poverty, 60–61
  teacher's challenges in, 106
Western curriculum in nonWestern classrooms, 122, 124–33
Western ethnocentrism
  anti-tradition tradition of, 40–41
  of high-level American politicians, 106–7
  ignoring nonWestern cultural knowledge, 9–10, 18, 28, 29
  imposing Western culture without respect for others, 26
  and objective intelligence measures, 47–48
  "our knowledge is superior" attitude, vii–viii, xi, 19–20, 21, 26, 53–54, 115
  in public schools and universities, 2, 71–73
  superiority of consumer-based society, 60
  superiority over nature, 82
  teachers' inability to recognize, 14–15
  See also high-status knowledge
Western public schools and universities, 2, 19, 71–73, 106–7, 127
Wilson, E. O., 50
wisdom of elders, 18
Woolfolk, Anita, 50
World Bank, 72–73, 87–88, 90–91, 100
World Trade Organization (WTO), 88–90, 96, 101, 122

Z

Zapotecs of Mexico, 69–70

other books by
C. A. Bowers

The Progressive Educator and the Depression: The Radical Years
Cultural Literacy for Freedom
The Promise of Theory: Education and the Politics of Cultural Change
The Cultural Dimensions of Educational Computing: Understanding the Non-Neutrality of Technology
Responsive Teaching: An Ecological Approach to Classroom Patterns of Language, Culture and Thought (with David Flinders)
Education, Cultural Myths, and the Ecological Crisis: Toward Deep Changes
Critical Essays on Education, Modernity, and the Recovery of the Ecological Imperative
Educating for an Ecologically Sustainable Culture: Rethinking Moral Education, Creativity, Intelligence, and Other Modern Orthodoxies
The Culture of Denial: Why the Environmental Movement Needs a Strategy for Reforming Universities and Public Schools
Let Them Eat Data: How Computers Affect Education, Cultural Diversity, and the Prospects of Ecological Sustainability
Educating for Eco-Justice and Community
Detras de la Apariencia: Hacia la Descolonizacion de la Educacion
Mindful Conservatism: Rethinking the Ideological and Educational Basis of an Ecologically Sustainable Future
Rethinking Freire: Globalization and the Environmental Crisis (edited with Frederique Apffel-Marglin)

A BOOK SERIES OF CURRICULUM STUDIES

This series employs research completed in various disciplines to construct textbooks that will enable public school teachers to reoccupy a vacated public domain—not simply as "consumers" of knowledge, but as active participants in a "complicated conversation" that they themselves will lead. In drawing promiscuously but critically from various academic disciplines and from popular culture, this series will attempt to create a conceptual montage for the teacher who understands that positionality as aspiring to reconstruct a "public" space. *Complicated Conversation* works to resuscitate the progressive project—an educational project in which self-realization and democratization are inevitably intertwined; its task as the new century begins is nothing less than the intellectual formation of a public sphere in education.

The series editor is:

>Dr. William F. Pinar
>Department of Curriculum and Instruction
>223 Peabody Hall
>Louisiana State University
>Baton Rouge, LA 70803-4728

To order other books in this series, please contact our Customer Service Department:

>(800) 770-LANG (within the U.S.)
>(212) 647-7706 (outside the U.S.)
>(212) 647-7707 FAX

Or browse online by series:

>www.peterlangusa.com